The Money Blueprint

A Step-by-Step System for Building Wealth from Zero

Joseph Trek

Published by NobleTrex Press

For permissions and other inquiries, write to:

P.O. Box 3132, Framingham, MA 01701, USA

Contents

Introduction

You are going to spend approximately 90,000 hours of your life working for money.

That is a staggering number. It represents decades of early mornings, late nights, stressful deadlines, and missed opportunities. You will trade your energy, your creativity, and your finite time on Earth in exchange for a digital currency that flows into your bank account.

Yet, despite this massive investment of life force, the vast majority of people spend less than ten hours learning how that money actually works.

We are taught how to solve for x in algebra, how to analyze literature, and how to recite the dates of historical battles. But we are sent into the world completely illiterate in the language of the very thing that dictates where we live, what we eat, and how much freedom we have. We are handed the keys to a high-performance vehicle-our financial potential-without a manual, a map, or even a basic understanding of how to drive.

The result is a paradox that defines modern life: we have high-income professionals living paycheck to paycheck, paralyzed by six-figure debt. We have smart, capable people who are terrified of the stock market because they equate it with a casino. We have a culture obsessed with looking rich, yet completely ignorant of how to *become* wealthy.

If you feel like you are working harder than ever but simply treading water, it is not because you are bad with numbers. It is not because you aren't smart enough. It is because you are playing a game where nobody told you the rules.

This book is the rulebook.

The Default Setting is "Broke"

There is a reason building wealth feels like swimming upstream. The entire modern economy is engineered to separate you from your money.

Every app on your phone, every algorithm on your social feed, and every credit card offer in your mailbox is designed by teams of behavioral psychologists and data scientists with one goal: to trigger your dopamine receptors and make spending money the path of least resistance. The default setting of modern life is consumption. If you do not have a plan, you become part of someone else's revenue stream.

Most financial advice tries to fight this with willpower. You are told to stop buying lattes, to clip coupons, and to feel guilty about enjoying your life. But relying on willpower is a strategy doomed to fail. Willpower is a battery that drains throughout the day; by the time you are tired and stressed on a Tuesday evening, the "smart" decision doesn't stand a chance against the easy one.

This blueprint offers a different approach. We are not going to rely on willpower. We are going to rely on architecture.

Wealth is not an accident, and it is rarely the result of a lucky lottery ticket or a crypto moonshot. Wealth is a structure. Like a skyscraper, it requires a foundation, a framework, and a maintenance system. If you build the structure correctly, it stays standing regardless of the weather. If you build it on sand, it collapses the moment the wind changes.

The Difference Between Income and Wealth

Before we lay the first brick, we must dismantle the most dangerous misconception in personal finance: the confusion between income and wealth.

Income is what you earn. It is the river of cash flowing into your life. Wealth is what you keep. It is the reservoir.

Society celebrates income. We ask, "What do they do for a living?" or whisper about someone's salary. But income is fleeting. If you stop working tomorrow, your income drops to zero. If you spend everything you make-whether that's $50,000 or $500,000-you are not wealthy; you are simply a high-transaction mechanism for the economy. You are a conduit, not a container.

True wealth is the accumulation of assets that generate value while you sleep. It is the point where your money begins to work harder than you do. This state of affairs-financial freedom-is not reserved for trust fund babies or Wall Street tycoons. It is a mathematical inevitability for anyone who follows the correct blueprint over a long enough timeline.

Engineering Your Escape Velocity

So, how do we move from the chaos of living paycheck to paycheck to the stability of wealth? We stop treating our finances like a mystery and start treating them like an engineering problem.

In the early pages of this book, we are going to look at the physics of money. We will examine the silent, relentless force of inflation-a phenomenon that is currently eating your savings at a rate that should alarm you. We will define your "Net Worth" not as a vanity metric, but as the only honest scoreboard you have. You cannot improve what you do not measure, and for many, simply calculating this number is the splash of cold water that wakes them up.

Once we understand the physics, we will build the engine. This is where most people get stuck. They try to budget using restrictive diets that make them miserable. Instead, we are going to design a "Cash Flow Engine." We will automate your financial life so that saving and investing happen in the background, without your active involvement. By the time you see your paycheck, the heavy lifting will already be done. You will learn to strip the emotion out of the transaction. When you automate good decisions, you remove the option to make bad ones.

But a good engine is useless if the car is weighed down by an anchor. That anchor is debt.

Debt is a product sold to you by banks that understand compound interest better than you do. When you owe money on a credit card, you are on the wrong side of the most powerful mathematical force in the universe. You are the source of someone else's passive income. We are going to implement a ruthless strategy to sever those chains. We will triage your liabilities, decode the credit score game, and position you to use credit as a tool for leverage rather than a shackle of servitude.

The Investor's Mindset

If saving is the defense, investing is the offense. This is the part of the blueprint that usually intimidates people the most. The financial industry has spent decades making investing sound complicated to justify their fees. They use jargon like "quantitative easing," "beta," and "derivative pricing" to make you feel small.

Here is the secret: Successful investing is incredibly boring.

If you are finding investing exciting, you are probably gambling. Real wealth is built by buying boring, productive assets and doing nothing for a very long time. We will strip away the noise of the 24-hour news cycle. You will learn why trying to "beat the market" is a fool's errand and why the most sophisticated investors in the world often choose the simplest strategies.

We will look at the stock market not as a squiggly line on a graph that goes up and down, but for what it actually is: a venue to buy ownership in the most profitable businesses in human history. We will construct a portfolio that lets you sleep at night, balancing the need for growth with the need for safety. You will learn to recognize the "fees" that act as termites in your woodwork, eating away half your future wealth if left unchecked.

The Ghost in the Machine

However, a blueprint is only as good as the builder. You can have the perfect mathematical strategy, the lowest-fee index funds, and the most optimized tax plan, and still ruin everything in a moment of panic.

That is because money is only 20% math. The other 80% is behavior.

We are biologically wired to be terrible investors. Our brains evolved to survive on the savannah, fleeing from predators and seeking immediate caloric gratification. We are wired to run when we see danger (selling when the market crashes) and herd together when we see safety (buying when the market is at an all-time high). This is the exact opposite of what builds wealth.

We will spend significant time rewiring your money mindset. We have to identify the cognitive biases-the mental glitches-that lead us to sabotage our own success. We will tackle the thorny, emotional decisions that don't fit into a spreadsheet, like whether to rent or buy a home (spoiler: owning a home is not always the gold standard of wealth), how to negotiate your salary, and how to protect yourself from the catastrophic risks that can wipe out decades of progress.

Your Timeline Starts Now

There is a Chinese proverb that says, "The best time to plant a tree was twenty years ago. The second best time is now."

Financial regret is a heavy burden. I meet people every day who say, "I wish I had started this when I was twenty-five." They look at the compounding charts and feel a pit in their stomach, calculating the millions they missed out on.

If that is you, put that baggage down. You cannot audit the past. You can only engineer the future.

The system in this book works whether you are twenty-two and landing your first job, or fifty-two and realizing you need to catch up. The mathematics of wealth are agnostic to your age, your background, or your past mistakes. They only care about what you do next.

This is not a get-rich-quick scheme. If you are looking for a hot stock tip or a way to double your money in a month, put this book down and go buy a lottery ticket. This is a get-wealthy-eventually system. It is slow, it is methodical, and it is inevitable.

We are about to take a chaotic, stressful, essential part of your life and bring it under your control. We are going to turn the lights on in the dark room of your finances.

You have worked hard for your money. It is time your money started returning the favor.

Turn the page. Let's get to work.

Chapter 1

The Foundation of Financial Freedom

Most people jump straight to 'What stock should I buy?' without checking if the floorboards of their financial house are rotten. This chapter isn't about boring theory; it's about the physics of money. We'll strip away the jargon to reveal the five unshakeable laws that determine whether you drift endlessly or build lasting wealth.

1.1 Defining Your Destination

A strange thing happens when people talk about money: the numbers start driving the conversation as if they were the point. "I want to be a millionaire." "I need to make six figures." "I should buy a house." Those sentences sound decisive, but they're often just borrowed slogans—financial bumper stickers that feel responsible without actually being specific.

Money isn't the destination. Money is fuel. Fuel is useless unless you know where you're going, and expensive if you drive in circles.

If you've ever opened a maps app and typed the wrong address, you know the feeling: the little blue dot is you, the route looks confident, and you're still headed somewhere you don't want to be. In personal finance, the "wrong address" usually comes from one of two places: (1) other people's expectations, or (2) your own vague anxiety dressed up as a goal. The fix isn't to learn a trick. It's to get brutally clear on what you're trying to *protect* and what you're trying to *unlock*.

The Two Questions That Reveal Your Real Goals

Most financial goals are just answers to questions nobody asked. Start with these two:

- What do I want my money to *do* for me?
- What do I refuse to trade away to get it?

That first question creates your target. The second creates your guardrails. Together, they turn "more money" into a plan you can actually follow on a random Tuesday when life is messy.

Here's the key idea: a good goal has *meaning*, and a good plan has *constraints*. People think constraints are limiting. In reality, constraints are liberating. They remove 90% of decisions you'd otherwise debate, second-guess, and procrastinate on.

If your goal is "retire early," you can rationalize anything: risky stock picks, extreme frugality, moving to a place you hate, skipping health care, burning out at work, or quietly abandoning the goal the first time you feel tired. But if your goal is "reach financial independence while still having dinner with my kids most nights," that constraint immediately filters out certain jobs, certain side hustles, and certain spending cuts. The goal becomes harder in one sense—you can't bulldoze your life to get there—and easier in another: you stop negotiating with yourself.

Values-Based Planning: Your Personal ``Why" Filter

Values-based planning isn't a motivational poster. It's a decision system. A value is not a preference ("I like nice hotels"). A value is a principle you keep even when it costs you something ("I prioritize rest and health," or "I prioritize generosity," or "I prioritize freedom of time over status").

Try this thought experiment: imagine you hit your number—the bank balance that's supposed to make you feel safe. Now imagine how you got it.

- If you hit the number by working 80-hour weeks for a decade and your relationships are brittle, was it worth it?
- If you hit the number by never traveling, never taking risks, never enjoying your life, does the victory feel hollow?
- If you hit the number by chasing a lifestyle you don't even want, does it still count as success?

Money can buy options. But it can also buy a very expensive version of the wrong life.

Values-based planning is how you prevent that. You write down the small set of outcomes that matter enough to steer everything else. Not ten. Not twenty. A handful. Three to five is plenty. Any more and you'll start contradicting yourself without noticing.

Common values that show up in real financial plans:

- *Stability:* sleeping well at night, predictable bills, fewer emergencies turning into crises.
- *Freedom:* the ability to say no—to a boss, a city, a schedule, an obligation.
- *Family:* time, presence, support, and sometimes the ability to help parents or kids.
- *Health:* food, care, rest, a life not designed around burnout.
- *Growth:* learning, creating, taking smart risks, building something.
- *Service:* giving, supporting a cause, contributing beyond yourself.

None of these are "better" than the others. The problem is pretending you can optimize all of them equally. You can't. Choosing values is choosing tradeoffs *on purpose* instead of letting them happen to you.

The Status Trap: Why ``Should" Goals Are So Sticky

If your financial goals feel heavy, there's a good chance they contain the word "should." You should own a home. You should upgrade your car. You should max out everything. You should never buy coffee. You should be grateful. You should hustle.

"Should" goals are sticky because they come with social rewards. They get nods. They get likes. They make you look responsible, ambitious, adult. But "should" goals often ignore the costs that don't show up on Instagram: stress, commuting time, maintenance, opportunity cost, and that constant low-grade feeling of living someone else's life.

A classic example is the house question. Owning can be a fantastic goal—*if* it matches your values and time horizon. But if your deepest value is flexibility (being able to relocate quickly for work, family, or adventure), a house can turn into an anchor. Not an evil anchor. Just a heavy one. Property taxes, repairs, selling costs, the mental load of maintaining it, the way it can make you say "no" to opportunities because moving is complicated.

Values-based planning doesn't answer "Should I buy?" with a universal rule. It answers with a better question: "What job is this house supposed to do in my life, and is there a cheaper way to get that same benefit?"

Sometimes the job is stability. Sometimes it's pride of ownership. Sometimes it's space for kids. Sometimes it's a belief that renting is "wasting money." That last one is usually the most expensive belief in the room because it turns a personal decision into a moral one.

Turn Values Into a Destination You Can Actually Aim At

A destination has to be specific enough that you can tell if you're moving toward it. But it also has to be human enough that you still care about it six months from now.

A useful format is:

> *I want a life outcome by a time frame, and I'm willing to trade these things but not these things.*

Notice what this does. It includes:

- an outcome (what your life looks like),
- a deadline (when it matters),
- and constraints (what you won't sacrifice).

Now translate that into financial targets. This is where a lot of people either get too abstract ("financial freedom") or too narrow ("save $50,000"). You want targets that create a system.

Think in three layers:

- *Values:* the principles (freedom, stability, family).
- *Goals:* measurable outcomes (emergency cushion, debt payoff, savings rate, ability to work part-time).
- *Tactics:* the day-to-day mechanics (automatic transfers, account choices, spending rules).

When these layers align, financial decisions start getting simpler. Not easy. Simpler. You stop having to reinvent your plan every time your coworker buys something shiny.

A Real-World Case Study: Two People, Same Income, Opposite Destinations

Consider Maya and Chris. They're both 29, both making $78,000, both smart, both tired of feeling like their money disappears. From the outside, they look identical. Their destinations could not be more different.

Maya's values: *freedom and exploration*. She wants the option to take a three-month break in two years to travel and do a volunteer project overseas. She doesn't need luxury; she needs time and mobility. Her non-negotiables: she refuses to carry high-interest debt and refuses to lock herself into a lifestyle that requires constant overtime.

Chris's values: *stability and family*. He wants to propose soon, buy a home in the same school district his partner loves, and have a baby within three years. He doesn't need maximum flexibility; he needs predictable housing and a cash cushion so a broken transmission doesn't become a crisis. His non-negotiables: he refuses to gamble with money needed in the next three years and refuses to build a plan that depends on a bonus.

If they both follow generic advice, they'll both get frustrated.

Generic advice for Maya might say: buy a house as soon as you can. But that would sabotage her value of mobility and create selling risk right when she wants to take a break. Her better targets might look like:

- Build a "Freedom Fund" of $18,000 within 24 months (enough to cover three months of core expenses plus the travel project).
- Keep monthly fixed costs low enough that a temporary income drop doesn't break her.
- Invest consistently for long-term wealth, but don't put near-term travel money at market risk.

Generic advice for Chris might say: invest aggressively because you're young. But his timeline is short and his risk capacity is different. His better targets might look like:

- Save $35,000 in 30 months for down payment and closing costs, in a safe place (because the date matters more than the return).
- Build an emergency fund that covers the realities of a future baby: medical deductibles, time off, surprise costs.
- Contribute enough to retirement to capture any employer match (free money), without derailing the home goal.

Same income. Same age. Completely different "best" plan.

This is what values-based planning fixes. It stops you from confusing other people's scoreboard with your own.

The Constraint Power Move: Decide Your ``Enough"

Most people don't overspend because they're careless. They overspend because they're undefined. When you don't know what "enough" looks like, every upgrade feels justified. Bigger apartment. Nicer car. Better brand. Faster shipping. More convenience.

"Enough" is not a poverty vow. It's a line you draw so your money stops leaking into status purchases you don't even remember a month later.

Here's a practical way to define it without becoming a monk: choose one or two categories where spending genuinely improves your life, and give yourself permission to spend there *on purpose*. Then choose one or two categories where spending is mostly autopilot or social pressure, and set hard constraints.

Examples:

- If you value health, maybe you spend on good groceries and a gym you actually use, but cap restaurants.
- If you value relationships, maybe you spend on hosting friends or visiting family, but drive a boring car.
- If you value freedom, maybe you keep housing modest, but spend on experiences.

Constraints aren't punishment. They're how you buy the future you said you wanted.

Make Your Goals ``Tradeoff-Proof"

Goals fail for predictable reasons:

- They're too vague to measure.
- They're too rigid to survive real life.
- They're based on a fantasy version of you.

A tradeoff-proof goal has two features:

- *It survives a bad month.* If one unexpected expense destroys your plan, it wasn't a plan; it was a streak.
- *It tells you what to do when goals conflict.* Because they will.

Conflicts are where values do the real work.

If you value stability more than speed, you build a bigger cash buffer before you take big investing risks.

If you value freedom more than image, you keep fixed costs low even if you can technically "afford" more.

If you value family more than optimization, you might choose a slightly lower-paying job with better hours, then adjust your savings targets rather than pretending you can do everything.

Financial advice becomes dangerous when it pretends trade-offs don't exist. Your life is not a spreadsheet; it's a set of competing priorities with deadlines and emotions attached.

A Practical Framework: The Three Time Horizons

Money decisions get clearer when you sort them by *when* you'll need the money, because time changes what "smart" looks like.

- **Short-term (0—2 years):** money you might need soon should prioritize safety and accessibility over growth. The job is not to impress you. The job is to be there.
- **Mid-term (2—10 years):** money for goals with a date but some flexibility. The job is balance: some growth, but not so much risk that a downturn wrecks the timing.
- **Long-term (10+ years):** money for future-you. The job is growth that outpaces inflation, which usually means accepting volatility along the way.

Values plug into this. If you value freedom, you may prioritize short-term liquidity. If you value legacy, you may prioritize long-term compounding. If you value stability, you may build a larger short-term buffer even if it slightly slows long-term growth. None of these is irrational. They're personal design choices.

Turn the Destination Into a ``One-Page Money Compass"

You don't need a 40-tab spreadsheet to act like a CEO of your finances. You need a compass you can consult when temptation shows up, when life changes, and when fear tries to put you into financial paralysis.

Write a single page (paper or digital) with these four lines:

- *My top values:* (3—5 words or phrases)
- *My 12-month outcomes:* (2—4 measurable targets)
- *My 3—5 year outcomes:* (1—3 measurable targets)
- *My non-negotiable constraints:* (what you won't trade away)

Then add one sentence: *"If I'm unsure, I choose the option that protects my constraints and moves me toward my 12-month outcomes."*

That sentence is the antidote to impulse spending and to goal-hopping. It also makes you faster. Most people think being good with money requires endless willpower. In reality, it requires fewer decisions, made earlier, when you're calm.

Actionable Takeaway: Do the 20-Minute ``Destination Draft" Today

Set a timer for 20 minutes and write rough answers—ugly is fine—to these prompts:

- "When I say *financial freedom*, I mean: ____"

- "If my life is working in three years, a normal Tuesday looks like: ____"
- "I am willing to trade ____________________ for that life, but I refuse to trade ____________________."
- "My top three values are: ____"
- "My most important measurable money targets for the next 12 months are: ______________________________"

Put that draft somewhere you'll actually see it—your notes app, your wallet, taped inside a closet door. The point isn't to get it perfect. The point is to stop letting default goals run your life, and start using money for what it's best at: buying stability, buying options, and buying time for the things you don't want to miss.

1.2 The Truth About Your Starting Point

People avoid checking their net worth for the same reason they avoid stepping on the scale after a long winter: it feels like inviting judgment into your living room. But net worth isn't a verdict. It's a reading. Like blood pressure. Like the gas gauge. Ignoring it doesn't protect you from reality—it just keeps you from steering.

The emotional trap is thinking a number equals a identity: "I'm good with money" or "I'm a mess." The practical truth is far less dramatic. Net worth is simply the snapshot that tells you whether your past choices are giving your future self a tailwind or a headwind. You can be a high earner with a negative net worth. You can be broke on paper but moving in the right direction. You can look "successful" and still be one flat tire away from chaos. The only way to know which story you're living is to measure.

Net Worth, in Plain English

Net worth is what you *own* minus what you *owe.*

Net worth = Assets – Liabilities

An *asset* is something you have that has value. A *liability* is a debt or obligation you have to pay.

That's it. No mystery. The confusion comes from two common mistakes:

- treating the calculation like a moral scorecard instead of a tool, and
- mixing up "stuff that costs money" with "things that are worth money."

A couch you love is valuable to your life. It is not necessarily an asset in your net worth. A car is often an asset (you could sell it), but it's also a machine that slowly turns into a smaller number. A college degree can be priceless, but it doesn't belong on the balance sheet because you can't sell it to pay rent. Net worth is not a measure of your worth; it's a measure of your financial *position.*

Why This One Number Changes Everything

A net worth statement does something a budget can't do: it shows whether you're building a foundation or building on sand.

Budgets are about *flow*—money in, money out. Net worth is about *stock*—the pile you're accumulating (or the hole you're digging). You can have a balanced budget and still be getting poorer in the ways that matter if debt is creeping up or savings are stagnating. You can also have a messy month-to-month budget and still be getting richer if you're steadily paying down high-interest debt or investing consistently.

Net worth gives you a different kind of clarity: it makes trade-offs visible. Suddenly, a decision isn't "Can I afford this?" It's "Is this purchase worth slowing down my escape velocity?" That question is quieter, but it's much harder to lie to.

The Personal Balance Sheet: Treat Your Finances Like a Business

Businesses don't guess. They keep a balance sheet. Not because they're cold or obsessive, but because reality is expensive when you ignore it.

Your personal balance sheet is a list with two columns:

- *Assets:* the things you own that have a dollar value
- *Liabilities:* the debts you owe

Then you subtract.

To make this painless, follow two rules:

1. Use rough numbers if exact numbers aren't immediately available.
2. Use today's value, not what you paid.

People get stuck trying to be precise. Precision is great later. The first version should be fast, honest, and done.

What Counts as an Asset (and What Doesn't)

Start with the obvious assets:

- cash in checking and savings
- retirement accounts (401(k), IRA, etc.)
- brokerage accounts
- the current market value of your home (if you own one)
- the current value of your car (if you own it)

Then the less obvious:

- HSA balance (if you have one)
- any cash-value life insurance (if applicable—many people don't have this, and many policies are not a good deal)
- business ownership value (if you own a business; keep it conservative)

Here's what usually *doesn't* belong:

- furniture, clothes, kitchen gadgets, hobby gear (unless it's genuinely valuable and realistically sellable, like certain collectibles)

- your "potential" income
- your education

A useful test is: *If I needed money in 30 days, could I sell this for something close to the number I'm writing down?* If the answer is no, keep it off. The goal is not to capture every object in your house. The goal is to understand your financial posture.

What Counts as a Liability (Debt That Actually Follows You Around)

List the debts:

- credit card balances
- student loans
- auto loans
- personal loans
- medical debt
- mortgage balance

Two details matter:

1. Write the *current* balance, not the original amount.
2. If you can, note the *interest rate* next to each debt, because it will later determine the order you attack them.

The interest rate note is important because debts are not all equal. A 3% student loan is a different animal than a 24% credit card. One is a slow leak. The other is a hole in the bottom of your boat.

The ``Net Worth Paradox": Why a Negative Number Isn't Always Bad News

If you're early in your career, you might calculate your net worth and see a negative number. That can feel like failure. It isn't automatically.

Negative net worth is common for people who:

- recently graduated (student loans)
- recently bought a car (auto loan)
- recently bought a home (mortgage)

A mortgage often creates a weird moment where you have a large asset (the home) and a large liability (the loan). The net worth number can be modest or even negative depending on down payment and closing costs. That doesn't mean buying a home was "wrong." It means you used leverage (borrowed money) to buy an asset, and leverage always looks dramatic on paper.

The key is whether the negative number is part of a deliberate plan with a reasonable payoff, or the result of drifting: credit card debt accumulating, payday loans cycling, balances growing with nothing lasting to show for it.

Net worth isn't a shame machine. It's a flashlight.

A Concrete Example: Jordan's ``Looks Fine" Life

Jordan is 34, earns $92,000, and feels "okay." Rent is paid. The car runs. Friends think Jordan is doing well. Jordan avoids checking accounts too often because it creates a tight feeling in the chest.

When Jordan finally does a net worth snapshot, here's what shows up:

Assets	
Checking + Savings	$3,200
401(k)	$18,500
Car (estimated resale value)	$14,000
Total Assets	$35,700

Liabilities	
Credit Card (22%)	$9,800
Student Loans (5%)	$21,500
Auto Loan (6%)	$16,200
Total Liabilities	$47,500

Net Worth	*–$11,800*

That number punches Jordan in the gut. But then something interesting happens: the shame starts to fade, replaced by something more useful—*specificity*.

Jordan can now see the problem isn't "I'm terrible with money." The problem is a profile:

- high-interest credit card debt is dominating the picture,
- there's no real cash buffer, and
- the car loan balance is higher than expected relative to the car's value.

Without the balance sheet, Jordan might try to "budget harder" in a vague way. With the balance sheet, Jordan can build a plan that makes structural changes:

- set up a starter emergency fund so a surprise expense doesn't go back on the card,
- prioritize the 22% debt aggressively because it's the financial equivalent of running uphill in sand,
- keep contributing enough to the 401(k) to capture any employer match (if available), because that's a rare kind of *instant return*,
- consider whether the car situation is worth refinancing, paying down, or changing in the future.

Jordan's life didn't change the moment the number appeared. Jordan's *options* changed. That's what measurement does: it creates choices.

The $400 Problem: Why ``Starting Point" Is Often About Shock Absorbers

A lot of financial plans collapse not because people don't know what to do, but because life keeps happening. The real enemy is not a lack of ambition; it's a lack of shock absorbers.

In the U.S., a meaningful share of adults report they would not cover a $400 emergency expense entirely with cash or its equivalent. That's not a trivia fact. That's a stress fact. It means a broken phone, a dental bill, or a car repair can force a chain reaction: credit card debt, missed payments, overdraft fees, maybe even missed work.

When you look at your starting point, don't just ask "What's my net worth?" Ask: *How fragile is my life to a normal-sized surprise?*

Fragility doesn't mean you're irresponsible. It often means you're human in an economy where costs jump and paychecks don't always cooperate. But it does change what you should do first.

How to Take the Snapshot Without Spiraling

People spiral because they turn the snapshot into a lifetime story. Avoid that by setting a few ground rules:

- You're measuring *today*, not your potential.
- You're measuring numbers, not character.
- You're building a baseline to improve, not a reason to panic.

If you feel dread, that's not a sign you should stop. It's a sign this measurement matters.

One practical trick: do the first draft in pencil (literal or metaphorical). It gives your brain permission to be imperfect and still finish.

A Simple Template You Can Copy

Create two columns. You can do this on paper, in a notes app, or in a spreadsheet. Here's the minimal version:

Assets	*Today's Value*
Cash (checking + savings)	
Retirement accounts	
Investments	
Car(s)	
Home value (if applicable)	
Other	
Total Assets	

Liabilities	*Balance Owed*
Credit cards	
Student loans	
Auto loans	
Personal loans	
Medical debt	
Mortgage (if applicable)	
Other	
Total Liabilities	

Net Worth (Assets — Liabilities)

If you don't know a number, write a reasonable estimate and put a star next to it. The stars become your to-do list for getting exact balances later.

A Quiet Upgrade: Add ``Liquidity" to the Snapshot

Net worth is the headline. Liquidity is the subtext.

Liquidity means how quickly you can turn something into cash without taking a big loss. A retirement account may be a strong asset but not very liquid (especially before retirement age). A home is valuable but not liquid. Cash is liquid. A credit card is not liquidity; it's borrowing.

Why does liquidity matter? Because liquidity is what keeps you from using debt as your emergency fund. When life hits, you don't want to be forced into the most expensive option simply because it's the fastest.

A simple liquidity check:

> *If I lost my income tomorrow, how many months could I cover my bare-bones expenses using cash (and true cash equivalents) without borrowing?*

Even if the answer is "two weeks," that's useful. It tells you the first layer of your plan isn't glamorous. It's shock absorption.

Credit Reports: The ``Other" Starting Point People Forget

Net worth tells you what you own and owe. Your credit reports tell lenders (and sometimes insurers, landlords, and employers in certain situations) what your financial history looks like.

Checking your credit reports isn't about obsessing over a score. It's about accuracy and defense. Errors happen. Fraud happens. If someone opens an account in your name, you want to know early, not when you're trying to rent an apartment or buy a car.

In the U.S., there is an official, federally authorized source for free credit reports: AnnualCreditReport.com. If you've never checked your reports, doing it once is like checking the locks on your doors. Most of the time, everything's fine. The point is making sure.

Actionable Takeaway: Do Your ``30-Minute Balance Sheet Sprint"

Set a timer for 30 minutes and create your first personal balance sheet. Don't research. Don't optimize. Just collect what you can quickly:

1. Write down your cash balances (checking + savings).
2. Pull up your retirement account balance(s) and write the totals.
3. List your debts and write the current balances (and interest rates if you can see them easily).
4. Estimate your car value and home value only if you can do it quickly; otherwise leave them starred for later.
5. Subtract liabilities from assets and circle the net worth number.

Then write one sentence underneath it, the sentence that turns a snapshot into motion:

> *"The one number I can improve this month is ______________________, and the first action I'll take is ________________________________."*

The goal isn't to feel good. The goal is to be *oriented*. When you know where you are, you stop relying on hope and start relying on leverage: the small, repeatable actions that make the numbers move in your favor.

1.3 Why Saving Alone Is Not Enough

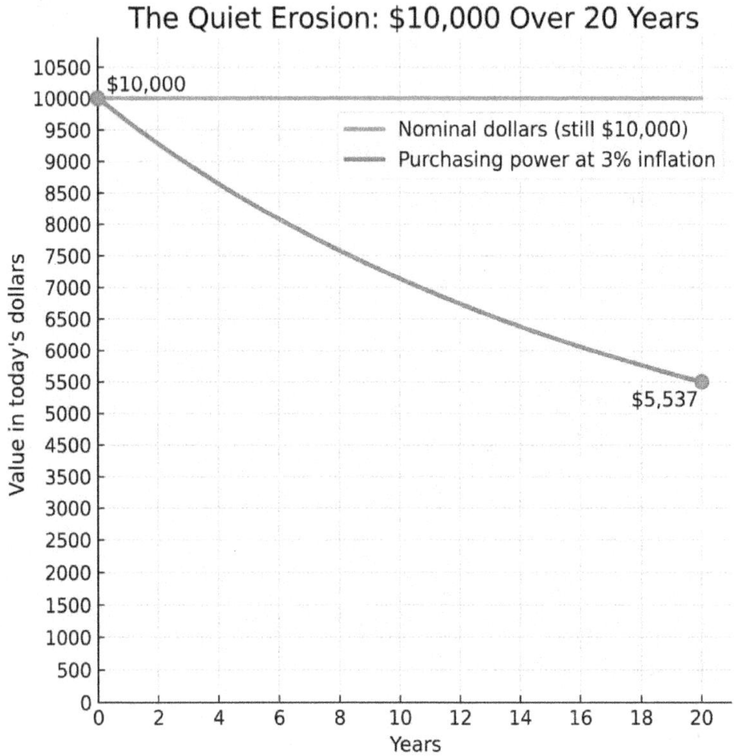

The most dangerous lie money tells is that it's sitting still.

If you put $10,000 in a drawer and come back in twenty years, you will find $10,000. Same digits. Same commas. Same satisfying weight when you imagine stacking it. And yet it will have quietly become a smaller life: fewer months of groceries, fewer plane tickets, fewer copays, fewer options. Inflation is that shrink ray. It doesn't kick in your door. It just reduces what your dollars can do, a little at a time, until you're forced to notice

People often talk about inflation like weather—unpleasant, out of your control, best ignored unless it's extreme. That attitude is expensive. Inflation is not a headline problem; it's a design problem. If your plan is "I'll just save cash," inflation turns your plan into a treadmill: you work, you save, you feel virtuous, and your purchasing power still drifts backward unless your savings are growing faster than prices.

The Difference Between ``More Dollars'' and ``More Life''

Here's the mental shift that changes everything: money has two versions of truth.

Nominal dollars are the simple dollars you see on a bank screen. $10,000 is $10,000.

Real dollars are what those dollars can buy. Real dollars are adjusted for inflation—for the fact that prices rise and your money's muscle weakens.

When people say "I got a 3% raise," their happiness depends on which truth they mean. If prices rose 4%, that raise is nominally positive and *really* a pay cut. If your savings account pays 1% while prices rise 3%, your balance will rise, but your buying power will fall. The bank app will congratulate you with a bigger number while your future self loses options.

That's why "I'm saving" is not automatically good news. Saving is a behavior. Wealth is an outcome. Inflation is the difference between the two.

Inflation Isn't Abstract. It's Tuesday.

Inflation is not "the economy." It's your grocery cart. It's your car insurance renewal. It's the price of a basic repair that used to be annoying and is now a small crisis. It's the daycare bill that makes parents feel like they're paying rent twice. It's the landlord's text that starts with "Due to increased costs..."

A plan that ignores inflation quietly assumes you'll outrun it with raises forever. Sometimes you will. Sometimes you won't. Careers pause. Industries change. Health happens. Babies happen. Burnout happens. The entire point of building wealth is to stop being at the mercy of events that don't ask your permission.

Inflation doesn't care whether you're responsible. It treats everyone the same: it charges rent on your idle cash.

The ``Safe" Feeling That Makes People Poorer

Cash feels safe for a reason. It doesn't bounce around. It doesn't send scary notifications. It doesn't drop 20% in a month. Cash is emotionally soothing.

But emotional safety and financial safety are not always the same thing.

Cash protects you from one kind of risk: *volatility* (your balance changing). But it exposes you to another: *purchasing power risk* (your life getting more expensive while your money stays still).

You can see this clearly when older relatives talk about prices from decades ago. "A movie ticket was $2." "My first apartment was $250 a month." They aren't exaggerating. They're describing the same phenomenon: the dollar used to lift more weight. That strength decays over time. If your money never grows, you're counting on yesterday's strength to pay tomorrow's bills.

A Concrete Case Study: Elena's ``Responsible'' Savings Plan

Elena is 27, cautious by nature, and proud of being disciplined. She keeps $18,000 in a savings account because it took real effort to build it, and she never wants to be broke again. She calls it her "security blanket."

On paper, Elena is doing the right thing. She has savings. She avoids debt. She doesn't impulse buy. Her friends call her the "adult" of the group.

Then she gets a promotion and starts thinking about bigger goals: a down payment, maybe kids someday, maybe the ability to switch jobs without panic. She looks at her savings and expects relief. Instead she feels something else: pressure. The number is the same number it was last year, but her life is getting more expensive. Her rent went up. Her insurance went up. Her grocery bill went up. Even her "simple" plans—a weekend trip, replacing a laptop—cost more than she mentally budgeted for.

Elena hasn't done anything wrong. She's simply discovered the limits of a cash-only strategy.

Here's the twist: the solution isn't to stop saving. The solution is to assign your dollars different jobs.

- Some dollars are *shock absorbers* (cash for emergencies and near-term needs).
- Some dollars are *builders* (money invested for long-term growth).

When every dollar is forced to be a shock absorber, you end up overpaying for safety. When every dollar is forced to be a builder, you end up underprotected and one surprise away from debt. A functional plan uses both.

Elena's breakthrough wasn't learning a new investment trick. It was realizing she needed a system: keep a cash buffer that actually lets her sleep, then send the rest to work somewhere inflation can't quietly mug it.

The ``Under the Mattress" Myth (and the Modern Version of It)

Nobody actually hides money under a mattress anymore, but the modern version is common:

- leaving large balances in checking accounts because it feels flexible,
- accumulating cash in low-yield accounts for years because investing feels intimidating,
- treating "not losing money" as the same thing as "getting ahead."

The trap is that cash has a visible risk (you can spend it), while inflation has an invisible one (it spends *you*). When a bank statement shows a stable number, your brain rewards you with a sense of control. Inflation is the opposite: it punishes you slowly, without a notification.

If you want a financial plan that survives adulthood, you need a way to see the invisible.

A simple way to visualize it is exactly what the chart at the top of the page shows: at 3% inflation, $10,000 loses roughly *half its purchasing power over a couple of decades*. Your nominal money stays the same; your real buying power slides downward like a bar of soap in a hot shower.

Why ``Just Save More'' Often Fails

When people feel stuck, they often reach for the most morally satisfying advice: "save more." It sounds clean. It sounds responsible. It also dodges the core issue.

Saving more is constrained by:

- income ceilings (your paycheck is your paycheck),
- cost floors (housing, food, healthcare don't negotiate much),
- and human limits (you can't optimize forever without hating your life).

A plan that requires you to squeeze harder every year is not a plan; it's a slow-motion burnout strategy.

Inflation makes "save more" a moving target. If prices rise, your savings goal rises too. If your money isn't growing, the finish line keeps walking away while you keep jogging.

The way out is not constant sacrifice. It's building an engine that can outpace inflation over time.

The Two Numbers That Matter: Inflation Rate vs. Your Return

Think of inflation as a competitor. It sets the pace. Your job is to beat it *after* costs, taxes, and fees.

If inflation averages 3% and your cash earns 0% to 1%, you're losing ground.

If inflation averages 3% and your investments average 6% to 8% over the long run (not every year—over time), you have a chance to grow purchasing power.

Notice the phrase *over time*. Investing is not a vending machine where you insert money and select "growth." Markets move. Some years are ugly. Some decades are weird. But the broad idea is unavoidable: if you want your future to be easier than your present, your money needs a way to grow faster than your costs.

This is where many people get stuck emotionally. "If investing is the answer, why does it feel so risky?" Because it is risky in the short term. The trick is matching the risk to the time horizon. Near-term money wants safety. Long-term money wants growth. Mixing them is what creates panic.

Inflation Risk Isn't Just a ``Cash" Problem

It's tempting to think inflation only punishes people who avoid investing. But inflation risk shows up inside investing too.

Even conservative, insured products can have inflation risk if their returns don't keep up with rising costs. You can "earn interest" and still get poorer in real terms. That's a painful sentence, but it's also clarifying: a positive number on a statement is not the same thing as progress.

This is why you don't evaluate financial decisions by whether they feel safe today. You evaluate them by whether they protect your purchasing power tomorrow.

The Fee Leak: How Investing Can Still Lose

Here's the second silent thief: fees.

Inflation reduces what your money can buy. Fees reduce how much your money can grow. The ugly part is that fees compound in reverse. A 1% annual fee doesn't sound like much. But over decades, it can carve away a shocking chunk of what you would have had, because it reduces the amount that gets to keep compounding.

This is where people get tricked by small numbers:

- "It's only a 1% expense ratio."
- "It's only a 1% advisory fee."
- "It's only a small plan fee."

Put them together and you can end up paying 2% or more each year—not once, but every year—for the privilege of getting market returns *minus* a permanent haircut. Meanwhile inflation is already taking its cut. If you're going to accept the discomfort of investing, you want as much of the return as possible staying in your pocket.

A solid plan respects a simple hierarchy:

> *First, don't get crushed by inflation. Second, don't donate your returns to unnecessary fees.*

So What Should Cash Actually Be For?

Cash is not the villain. Cash is a tool. The problem is when cash is forced to do a job it cannot do.

Cash is excellent for:

- emergencies (the car repair that can't wait for a market rebound),
- near-term goals (money needed in the next year or two),
- smoothing life (preventing overdrafts, late fees, and the stress of timing bills).

Cash is terrible for:

- long-term wealth building,
- retirement funding over decades,
- goals that require purchasing power to grow over time.

A lot of financial anxiety is simply misassigned money. Someone keeps too little cash and feels constantly at risk. Someone keeps too much cash and feels constantly behind. The right balance depends on your life, but the principle stays the same: match the money to the job.

The ``Treadmill" Test: Are You Building or Just Maintaining?

Here's a quick diagnostic you can do without a spreadsheet.

Imagine you froze your savings behavior exactly as it is today. Same account. Same contributions. Same habits. Now fast-forward five years. Would your money be buying you *more freedom*, or would it simply be keeping you in place while costs rise?

If the answer is "keeping me in place," you're on the treadmill. Treadmills are fine for workouts. They are terrible for building wealth.

Getting off the treadmill usually doesn't require extreme change. It requires one structural decision: separating your plan into short-term safety and long-term growth, so the money that needs to be safe stays safe, and the money that needs to grow gets a chance to grow.

Actionable Takeaway: The ``Two-Bucket Assignment''

Take 25 minutes and assign every dollar you save into one of two buckets. Not physically—mentally and in account structure if possible.

1. *Safety Bucket (Cash):* Decide what this bucket is protecting you from. Write the number you want in it based on your life (rent/mortgage, deductibles, job stability, responsibilities). If you're not sure, start with a modest target that would prevent a $400–$1,000 surprise from turning into credit card debt, then build from there.
2. *Growth Bucket (Investing):* Decide what this bucket is buying: retirement options, the ability to reduce work later, long-term freedom. This bucket exists specifically to fight inflation over decades.

Then write one rule you will actually follow:

> *"When my Safety Bucket reaches $________, every new dollar I save goes to the Growth Bucket automatically."*

That single sentence turns "I should invest someday" into a system. It lets you keep the comfort of cash *where it's appropriate* while giving your long-term money a way to stop shrinking in real terms. The result isn't just a bigger number; it's a future where your savings buys more life, not less.

1.4 The Engine of Wealth Creation

Picture two people doing the exact same responsible thing: setting money aside every month.

One checks the balance and sees a steady, comforting climb. The other checks the balance and sees something that looks almost broken at first: slow growth, then a bend upward, then a curve that starts to feel like the account is growing a life of its own.

The difference isn't hustle. It isn't a secret stock tip. It's a law of math that doesn't care how you feel about it: *money can earn returns, and those returns can earn returns*. That second layer—returns on returns—is the engine.

Most people don't struggle with the concept of saving. They struggle with waiting long enough for compounding to become visible. Compounding is like watching a kettle heat: for a while, it looks like nothing is happening. Then it changes state.

Compounding, Without the Finance Fog

Compound interest (or compound growth) means you earn money not just on what you put in, but also on the gains you already earned.

If you invest $1,000 and it grows by 7% in a year, you end the year with $1,070. If it grows by 7% again, you don't earn $70 the second year—you earn about $74.90, because you're earning on $1,070 now, not $1,000. That difference looks tiny. Over decades, it becomes the whole story.

Simple interest is like getting paid only on your original deposit. Compound growth is like getting paid on your deposit *and* your past wins.

Compounding doesn't require genius. It requires time, consistency, and not sabotaging the engine with leaks (fees, taxes you could legally reduce, panic-selling, constantly starting over).

Why Time Beats Talent (and Sometimes Beats Income)

Time is the one advantage you can't fake later. You can earn more later. You can learn more later. You can tighten spending later. But you cannot go back and buy ten extra years of compounding.

That's why the most frustrating sentence in personal finance is also the truest: starting early with less can beat starting late with more.

It feels unfair until you remember what compounding is doing. It's not just adding new contributions. It's multiplying old ones.

A useful mental model: your contributions are the seeds; compounding is the growing season. Planting more seeds matters. But a longer growing season is a different kind of advantage.

The Rule of 72: A Back-of-the-Napkin X-Ray

If you want a quick way to understand compounding without pulling out a spreadsheet, use the *Rule of 72*.

$$\text{Approximate years to double} \approx \frac{72}{\text{annual rate of return (in \%)}}$$

At 6% growth, money doubles in about $72/6 = 12$ years. At 8% growth, money doubles in about $72/8 = 9$ years.

This rule is not perfect, but it's good enough for intuition—and intuition is what keeps you from making expensive emotional decisions when markets get scary.

The deeper lesson is the punchline: doubling isn't linear. It doesn't happen smoothly. It happens because enough time passes for the math to stack on itself.

A Real-World Case Study: Same Saver, Different Start Dates

Meet Sam and Priya. They're both steady, not flashy. Neither is trying to "get rich quick." They just want options.

Sam starts investing at 25 and invests $300 a month until 35, then stops completely. Life gets busy. Kids. Mortgage. The money stays invested, but no new contributions.

Priya starts at 35 and invests $300 a month until 65. She invests for thirty years, three times as long as Sam contributed.

Here's the surprise many people hate: depending on the return, Sam can end up *very close* to Priya—and sometimes ahead—even though Sam invested for only ten years. Why? Because Sam's money had forty years to compound (from 25 to 65), while Priya's had thirty.

This isn't an argument for stopping at 35. It's an argument for respecting time. The earliest dollars you invest can become the most powerful dollars you ever invest, because they have the longest runway to multiply.

Priya's win is consistency. Sam's win is time. The best plan is both: start early *and* keep going.

The Emotional Barrier: Why Compounding Feels Fake at First

Compounding is mathematically real and psychologically weird.

Early on, your account growth is dominated by your contributions. You put in $300, the balance goes up about $300. The engine is running, but it's not the headline.

Later, your growth is dominated by the engine. The account can grow by more than your annual contributions. That's when people start calling it "passive income" and acting like it's magic. It isn't magic. It's just delayed visibility.

This delay is why so many people quit right before the payoff. They invest for a year or two, the balance doesn't look impressive, and they decide investing "doesn't work." That's like quitting a gym after three workouts because you don't have abs yet. The system isn't broken. The calendar is.

The Snowball Effect (and the Part Nobody Mentions)

Compounding is often explained as a snowball rolling downhill, picking up snow. The metaphor is fine. The missing detail is more important: snowballs melt when you keep taking bites out of them.

The biggest bites usually look like:

- high fees,
- taxes you could have avoided with better account choices,
- stopping and starting (selling when scared, buying back later),
- and constantly chasing "the next thing" instead of letting a simple strategy run.

Compounding is powerful, but it's not indestructible. Treat it like an engine: it needs fuel (contributions), time (runway), and maintenance (keeping costs low and behavior steady).

The Picture Your Brain Needs: Linear Saving vs. Compounding

Most people *understand* compounding only after they *see* it. The curve is the lesson.

Below is a simple comparison:

- A *linear saver* contributes consistently but earns no return (money sits in cash).

- A *compound investor* contributes the same amount but earns a steady average return over time.

The distance between the two lines is not "luck." It's what happens when your money is allowed to do work while you sleep.

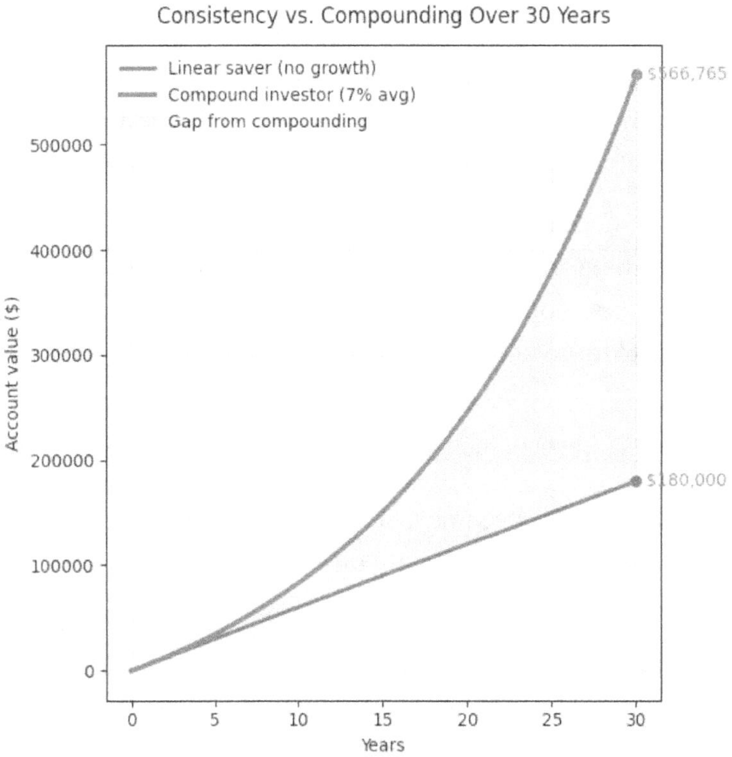

If the chart makes you feel impatient, that's normal. The curve is a reminder that the early years are supposed to look underwhelming. The "hockey stick" doesn't kick up because you finally became disciplined; it kicks up because enough prior money accumulated to generate meaningful growth.

The Hidden Superpower: Automating the Boring Part

The most reliable way to capture compounding is to make it boring.

Automation turns investing from a monthly willpower contest into a background process. When contributions happen automatically, you stop debating with yourself. You also avoid a subtle trap: waiting for the perfect moment. The perfect moment is rare and usually obvious only in hindsight. Consistency beats timing for almost everyone.

If you do nothing else, do this: set an automatic contribution amount that won't wreck your life, then increase it when your income rises. Compounding loves raises because they let you contribute more without feeling deprived.

Account Choices: When Taxes Act Like a Reverse Fee

You don't need to become a tax expert to understand the basic point: taxes can reduce what you keep, and what you keep is what compounds.

Some accounts are designed to give your compounding engine better conditions—less drag from taxes now, later, or both (depending on the account type and your situation). The right choice depends on your income, benefits, and goals, but the principle stays the same: *the less friction on your returns, the more your money can multiply.*

This is the part of wealth-building that feels unfair in a useful way: the system often rewards people who set up the right containers and then simply keep feeding them.

Fees: The Engine Killer That Wears a Suit

If compounding is the engine, fees are sand in the gears.

A fee isn't just money you pay. It's money you pay *plus all the growth that money could have generated for the rest of your life.* That's why small percentages matter so much. A 1% difference in annual cost can mean a dramatically different ending balance over decades, even when everything else is identical.

This is one reason low-cost, diversified investing tends to win for ordinary people: you're not trying to be the smartest person in the market; you're trying to keep more of the market's return.

A practical way to think about it:

> Every recurring fee is a permanent partner in your portfolio. Choose partners carefully.

What About Active Investing and Picking Winners?

It's tempting to believe wealth is built by being right—finding the next big stock, the next hot fund, the next perfect entry point.

Sometimes people do get lucky. But luck is not a system. A system is what still works when you're tired, busy, stressed, and not in the mood to research.

The compounding engine doesn't require you to predict. It requires you to participate in growth with discipline, keep costs low, and stay invested long enough for the curve to bend.

If you want to experiment with picking stocks, do it with a small, clearly labeled "fun money" slice that won't blow up your destination. Keep the core boring. Boring is underrated; boring pays rent.

Actionable Takeaway: Build Your ``Compounding Protocol" in 30 Minutes

Set a timer and write three lines, then act on them immediately:

1. *My long-term investing amount:* "I will invest $______ per month automatically."
2. *My annual increase rule:* "Each time my income rises, I increase that amount by ______% (or $______)."
3. *My fee ceiling:* "For my core investments, I will avoid high ongoing costs and keep total recurring fees as low as reasonably possible."

Then schedule the automation—not "soon," not "when you've done more research," but on a calendar date you can point to. Compounding rewards action that feels almost too small to matter. A monthly transfer you barely notice today can become the loudest financial decision you ever made, because time will keep amplifying it long after the motivation fades.

1.5 Understanding Risk and Reward

Most people say they want higher returns the way they say they want to be in shape: as a general preference, not as a commitment to the process. The moment the process gets uncomfortable—a market drop, a scary headline, a friend bragging about a "sure thing"—the brain tries to fire the emergency exit. That emergency exit has a name: "I'll just stay safe."

The problem is that "safe" is not one thing. Some choices feel safe because they don't wobble, but they quietly erode purchasing power. Other choices wobble a lot but, over long stretches of time, have historically been the main way ordinary people built wealth without winning the lottery. Risk isn't a monster under the bed. It's a menu of tradeoffs. If you don't choose your risks, you still get them—you just get the ones you didn't mean to order.

Risk Isn't the Enemy. Unplanned Risk Is.

The goal isn't to avoid risk. The goal is to take the risks you can afford, for the rewards you actually need, while refusing the risks that can permanently damage your life.

That sentence is doing a lot of work. Notice the three parts:

- *Risks you can afford* (your financial shock absorbers matter).
- *Rewards you need* (your destination matters).
- *Refusing permanent damage* (some risks are simply not worth it).

When people get burned in investing, it's rarely because they were "too greedy." It's usually because they took risks that were mismatched to their reality: investing money needed for rent, loading up on one stock because it felt exciting, or panicking out of the market at the worst possible time because they didn't have a cash buffer.

A smart plan doesn't eliminate fear. It designs around it.

Volatility vs. Permanent Loss: A Bumpy Ride and a Cliff

To make risk feel less mystical, separate it into two categories.

Volatility is the bumpy ride: prices go up and down, sometimes violently, and your account balance swings around. Volatility is unsettling, but it's not inherently fatal. If you stay on the ride long enough, volatility can be the price you pay for growth.

Permanent loss is the cliff: the money doesn't come back because it was concentrated in something fragile (a single company that fails), you sold at the bottom and never got back in, or you took on a debt spiral that forced you to liquidate at the worst time.

People often confuse these. They see volatility and assume permanent loss. They treat any drop as a signal of danger, when it may simply be the normal turbulence of markets doing what markets do.

The practical difference is simple:

> Volatility hurts your feelings. Permanent loss hurts your future.

You can design around feelings. You must avoid cliffs.

The Sneakiest Risk: Inflation Risk

It's tempting to think risk only exists in stocks. That's convenient because it means you can "avoid risk" by avoiding the market. But inflation doesn't give you that option.

When prices rise faster than your cash grows, you take inflation risk. It's a real risk with real consequences: delayed retirement, working longer than you want, being forced to cut back in old age, or feeling financially squeezed even when you're "doing everything right."

Inflation risk is also emotionally sneaky because it doesn't create dramatic down days. It creates quiet down years. Your money looks stable. Your life gets more expensive. The damage shows up as opportunities you can't take.

Once you see inflation as a risk, investing stops being a luxury activity for rich people and becomes what it really is for everyone else: a tool for defending your future purchasing power.

The Three Questions That Set Your Risk Level

You don't pick a "risk tolerance" like you pick a favorite color. Your risk level is shaped by your life.

Ask three questions:

- *Time horizon:* When do I need this money?
- *Risk capacity:* If this account dropped 30% tomorrow, would my life break?
- *Risk tolerance:* If this account dropped 30% tomorrow, would I panic-sell?

Time horizon is objective. If you need the money in two years for a down payment, your portfolio should not be designed like you're investing for retirement in forty years. Risk capacity is also mostly objective: it depends on your cash buffer, job stability, and obligations. Risk tolerance is the psychological piece: what you can emotionally endure without making a destructive decision.

Most people focus only on risk tolerance—"I'm conservative" or "I'm aggressive"—and ignore capacity and time. That's how you get someone who calls themselves conservative but has no emergency fund and is secretly taking massive risk because any market drop forces them into credit card debt.

A Concrete Case Study: The Same Market Drop, Two Different Outcomes

Consider Alex and Renee. Both invest $10,000 in a diversified stock index fund. A year later, the market falls sharply. Their accounts drop to $7,000.

Same math. Different outcome.

Alex has three months of expenses in cash and stable income. Alex feels nervous, complains about the news, and keeps contributing. Alex experiences volatility—the bumpy ride—but doesn't face a cliff.

Renee has $400 in checking and is already juggling bills. A car repair hits the same month the market drops. Renee sells the investment at $7,000 to cover the repair and a couple of late bills. Renee experiences volatility *and* gets forced into permanent damage: selling low because there was no cash buffer.

This is why "investing advice" without life context is irresponsible. It's not enough to tell people to invest. The real question is: *Can you stay invested when the world gets loud?* The answer depends less on bravery and more on preparation.

The unglamorous truth: emergency savings are not just a safety tool. They are an investing tool. They prevent forced selling.

Asset Allocation: The Main Dial You Control

If you only learn one technical idea about managing investing risk, make it this: *asset allocation*.

Asset allocation is simply how you split your money among different types of investments—most commonly stocks and bonds (and sometimes cash). It is the risk dial. Not the stock you pick. Not the headline you react to. The overall mix.

In plain language:

- *Stocks* tend to be bumpier but have historically offered higher long-term growth.
- *Bonds* tend to be steadier (not risk-free) and often reduce the stomach-churning swings.
- *Cash* is stable in nominal terms, useful for near-term needs, but vulnerable to inflation over long periods.

Your allocation should match your time horizon and your capacity to handle swings. The longer the horizon, the more room you typically have for volatility. The shorter the horizon, the more your plan should prioritize stability, because you don't have time to wait for a recovery.

Diversification: The ``Don't Bet Your Life on One Thing" Rule

Diversification means spreading your investments so one bad outcome doesn't wreck you.

It's not about owning a dozen random stocks. It's about reducing concentration risk. If one company implodes, or one sector struggles for years, diversification keeps your plan from depending on a single storyline.

This is how wealth gets built by ordinary people: not by being right all the time, but by refusing to be catastrophically wrong.

A vivid way to think about diversification:

> You don't need every investment to win. You need no single investment to be able to end you.

Diversification doesn't eliminate losses. It reduces the chance of a permanent, unrecoverable one.

Rebalancing: The Boring Habit That Prevents Accidental Risk

Over time, your portfolio drifts. If stocks do well for a few years, they become a larger share of your portfolio. You become "more aggressive" without choosing it. If stocks crash, you become "more conservative" without choosing it.

Rebalancing is the practice of bringing your portfolio back to its target mix. It's a discipline tool. It forces you to do the emotionally difficult thing in a controlled way:

- sell a little of what's gone up,
- buy a little of what's gone down,

- and return to your intended risk level.

This is not about perfection. It's about preventing drift from turning into accidental, unplanned risk.

The Seduction of ``High Return'' Promises

Whenever markets get uncertain, "guaranteed" opportunities multiply like fruit flies.

Someone has a strategy that "can't lose." Someone has a real estate deal that "always cash flows." Someone has a crypto project, an options system, a private fund, a friend-of-a-friend who "knows what's about to happen."

The common thread is not that these are always scams (though many are). The common thread is that they offer relief from uncertainty. They sell the feeling of control.

But risk and reward are linked for a reason. Higher expected return generally requires accepting more uncertainty. When someone offers high returns with no risk, what they're usually offering is hidden risk: leverage, illiquidity (you can't get your money back), concentration, or outright fraud.

A mature money mindset doesn't ask, "Is this safe?" It asks, *"Where is the risk hiding?"*

Fees and Risk: Paying More Doesn't Make It Safer

There's a comforting myth that expensive investing is safer investing: higher-fee funds, complicated products, constant trading, fancy strategies, a sense that someone is "doing something."

Fees are guaranteed. Returns are not.

If you take more market risk to chase higher returns but leak 1–2% per year in unnecessary costs, you've increased your uncertainty without fully capturing the reward. Over decades, that drag can be enormous. Fees are a form of permanent loss that masquerades as sophistication.

Low-cost doesn't mean low-quality. In investing, low-cost often *is* the quality feature, because it preserves the engine you're trying to build.

Designing a Risk Plan You Can Live With

A risk plan has to survive two versions of you:

- *Calm you*, reading about investing on a quiet evening.
- *Stressed you*, checking the news after a bad market day.

Calm you makes rational decisions. Stressed you makes protective decisions. A good plan respects stressed you by building guardrails:

- keep an emergency fund so you're not forced to sell,
- don't invest near-term goal money in volatile assets,
- choose a diversified approach so one thing can't wreck you,
- automate contributions so you don't rely on mood,
- and set a simple rebalancing routine (calendar-based is fine).

This is what "risk management" looks like in real life. It's not a spreadsheet trick. It's building a structure strong enough to hold your behavior when fear shows up.

Actionable Takeaway: Write Your ``Risk Rules" Before You Need Them

Take 15 minutes and write a one-paragraph risk policy you'll follow even when you're nervous. Put it somewhere you can find it.

Here's a template you can copy and fill in:

> *"Money I need within ______ years stays in cash (or low-volatility options). My emergency fund target is $________, so I am never forced to sell investments to handle normal surprises. For long-term goals (10+ years), I accept volatility as the price of growth and use a diversified, low-cost approach. I will not make big changes based on headlines. If my portfolio drifts, I rebalance ______ times per year. If the market drops sharply, I will continue my automated contributions and revisit my plan only after ______ days."*

The point isn't to predict the future. The point is to remove the most dangerous version of risk: the kind created by panic, concentration, and improvisation. When your rules are written ahead of time, volatility becomes a bumpy ride you chose—not a cliff you stumbled off.

Chapter 2

Building Your Cash Flow Engine

Most financial advice focuses on net worth-a static number on a spreadsheet. But life happens in motion. You cannot pay for groceries with home equity, and you cannot invest money you have already spent. This chapter is about building your 'Cash Flow Engine,' a set of automated pipes and buckets that captures your income, directs it exactly where it needs to go, and funds your freedom before you even have the chance to spend it. We are moving from relying on willpower to building a machine that works for you while you sleep.

2.1 Mastering Money in Motion

Net worth is a photograph. Cash flow is a movie.

A photograph can be gorgeous and still hide trouble. Someone can own a nice car, have a decent-looking apartment, even hold a retirement account they never think about—and still feel like money evaporates the second it arrives. The problem isn't always *how much* you have. It's how money *moves* through your life: when it shows up, where it leaks out, and whether any of it ever gets captured.

Cash flow is simply the map of *money in* (income) and *money out* (expenses) over time. If you learn to read that map, you stop guessing. You stop blaming yourself for being "bad with money" when the real issue is a system with holes in it. And you start building wealth the way it's actually built in the beginning: not by picking perfect investments, but by creating a reliable gap between what you earn and what you spend—then protecting that gap like it's oxygen.

Here's the small but life-changing shift: stop asking, "How much money do I have?" and start asking, "How much money do I keep *each month*?"

Cash flow has a quiet superpower: it reveals cause and effect. You can't negotiate with a credit card statement, but you *can* spot patterns. You can see that your "random" spending isn't random at all; it clusters around weekends, or stressful workdays, or the week before payday. You can see that the money stress you feel is less about totals and more about timing—rent hits on the 1st, your paycheck lands on the 5th, and those four days feel like walking a tightrope over overdraft fees.

And once you see the motion, you can design it.

The Tank, the Drains, and the Bucket

Imagine your finances as a water system.

Income is the faucet filling a tank. Your tank is your checking account—the place money gathers before it runs out into the world. Then there are drains. Some are unavoidable and predictable: housing, utilities, insurance, minimum debt payments. Others are flexible: restaurants, rideshares, clothes, hobbies, gifts, the *"it's been a long week"* purchases.

Savings is what happens when the tank fills high enough that water spills over into a separate bucket you set aside on purpose. If the drains are too wide, the water level never reaches the overflow point. You're not failing; the plumbing is.

Right after you internalize that idea, most money decisions get less emotional. You're no longer debating whether you *deserve* to save. You're asking a calmer, more practical question: *Which drain do I narrow first so the overflow finally happens?*

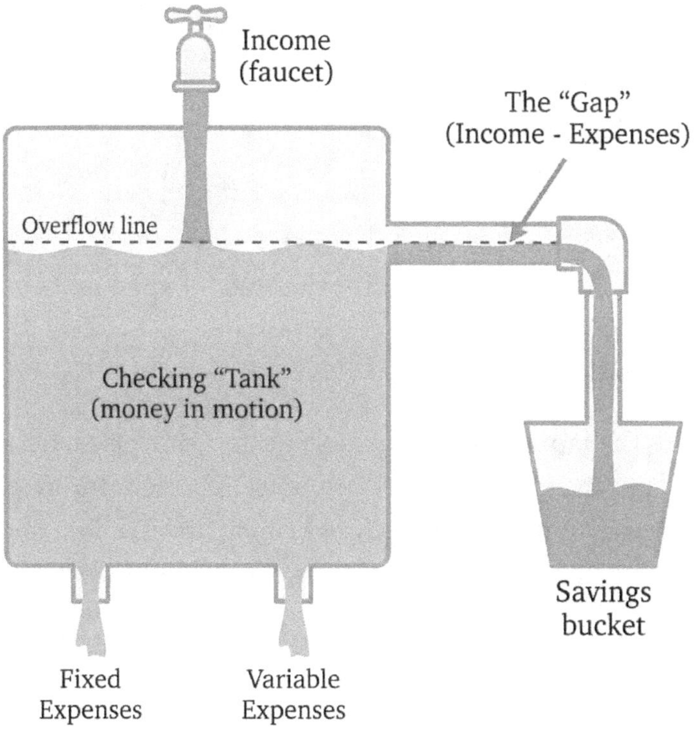

Why ``More Money" Doesn't Fix a Leaky System

There's a cruel little pattern in personal finance: when income rises, spending often rises to match it. Not because people are reckless, but because life expands to fill the container you give it. A slightly nicer apartment feels reasonable. Convenience subscriptions multiply quietly. Delivery becomes a habit. A higher car payment arrives dressed up as a reward for hard work.

If your drains widen as fast as your faucet, the tank never overflows. You can make more and still save nothing.

That's why cash flow, not salary, is the early wealth-builder's obsession. Savings isn't a personality trait. It's math plus design.

A useful mental model is this:

$$\text{Savings (per month)} = \text{Income} - \text{Expenses}$$

That's the gap. The gap is your engine. Investments matter, but early on, the gap is the fuel you pour into every future goal: emergency fund, debt payoff, retirement, the freedom to quit a bad job, the ability to move cities, the power to say no.

When people say they want to be "financially free," they usually mean they want their future self to have options. Options are purchased with cash flow.

Your Personal Cash Flow Statement (Not a Budget, a Truth Serum)

Forget perfection. Forget tracking every penny forever. Think of a Personal Cash Flow statement as a simple scoreboard. It answers three questions:

1. How much money reliably comes in each month?
2. Where does it reliably go?
3. What's left—and is that "leftover" real, or just an illusion created by delaying bills?

A shockingly common experience: someone checks their account on a random Tuesday and feels rich. Then the 1st of the month hits and three autopays punch the account in the throat. That's not a moral failure. It's a visibility failure.

Start with the simplest version of the statement:

- *Monthly income*: net pay (after tax) plus any regular side income.
- *Fixed expenses*: the bills that tend to show up no matter what.
- *Variable spending*: everything else.
- *Margin*: the gap (income minus total spending).

That margin is what you're trying to grow. Not with suffering. With smart plumbing.

Burn Rate: The Number That Tells You How Fragile Life Is

Your burn rate is what it costs to keep your life running at its current baseline. It's the amount you *must* pay each month to avoid immediate consequences.

Burn rate isn't only rent and utilities. It includes anything with a penalty attached if you miss it: minimum debt payments, insurance, phone bill if your job depends on it, childcare, transit required to earn income.

Why it matters: burn rate tells you how much flexibility you actually have. Two people can earn the same income and have completely different financial stress levels because their burn rates are different.

Lower burn rate creates breathing room. Breathing room creates better decisions. Better decisions create wealth.

Here's a quick way to think about it: burn rate is your financial "idle speed." A high idle speed means the engine is revving even when you're parked. You can still drive fast, but you're burning fuel just sitting there.

Case Study: Two Paychecks, Same Salary, Opposite Outcomes

Consider Maya and Chris. Both earn $60,000 a year gross. After taxes and benefits, each takes home roughly $3,700 a month. Same faucet.

Maya's fixed expenses look like this:

- Rent + renters insurance: $1,450
- Utilities + internet: $220

- Car payment + insurance: $430
- Minimum student loan payment: $220
- Phone: $70
- Subscriptions (a few): $40

Her fixed total is about $2,430. She spends about $900 on groceries, gas, eating out, and fun. That leaves a margin of about $370.

Chris's fixed expenses look like this:

- Rent in a newer building: $1,900
- Utilities + internet: $260
- Car payment + insurance (newer car): $650
- Minimum student loan payment: $220
- Phone: $90
- Subscriptions (many): $120

Fixed total: $3,240. He spends about $700 variable. Margin: −$240. The gap is negative, so the system steals from the future each month—credit cards, BNPL, overdrafts, "I'll catch up next paycheck."

Neither person is "better." They just built different plumbing.

Now comes the interesting part. Maya doesn't feel wealthy, because $370 a month isn't glamorous. But that margin is a seed. If she simply routes that gap into savings automatically, she's building an emergency fund without drama. Chris, on the same income, feels constant pressure because there is no overflow. The tank never gets ahead of the drains.

One decision often creates the whole difference: housing and transportation. Those two drains are huge, and they're the drains society pressures you to widen as a symbol of adulthood. Cash flow doesn't care about symbols.

The True Savings Rate: Your Wealth-Builder Metric

Savings rate is the percentage of your income you keep, not the dollar amount. It turns cash flow into a clean, comparable metric.

$$\text{Savings Rate} = \frac{\text{Savings}}{\text{Income}} \times 100\%$$

If you take home \$3,700 a month and save \$370, that's a 10% savings rate. If you save \$740, that's 20%. This number matters because it predicts how quickly you can build an emergency fund, pay off debt, and start investing.

Early on, a higher savings rate is often more powerful than chasing investment returns. A 1% difference in your investment performance is nice. A 10% difference in your savings rate changes your life.

And savings rate is not the same as "whatever is left at the end of the month." Leftover money tends to get assigned a job by someone else—a friend's birthday dinner, a flash sale, a car repair, an app that makes spending frictionless. Savings rate means you decide first.

The Two Levers: Increase the Faucet or Narrow the Drains

There are only two ways to create overflow:

1. Increase income (turn up the faucet).
2. Reduce expenses (narrow the drains).

Most advice gets stuck in a false argument: "You can't cut your way to wealth" versus "Just stop buying lattes." Both miss the point. Cash flow is a system; you tune it.

Income growth is powerful, but it often takes time. Expense reduction can be immediate, but it has limits. The best approach is to do the *highest-impact, least-miserable* changes first.

A surprisingly effective rule: focus on the big, boring drains before the small, emotional ones.

- Big boring drains: housing, transportation, insurance, debt interest, recurring subscriptions.
- Small emotional drains: coffee, snacks, little conveniences that make life feel nicer.

Cutting every small joy can make a plan collapse out of resentment. But narrowing one big drain can create overflow while leaving your daily life mostly intact.

For example: negotiating car insurance, refinancing high-interest debt, moving when a lease ends, taking on a roommate temporarily, or trading a new car payment for a reliable used car. These aren't glamorous, but they're structural.

Cash Flow Timing: When Money Moves Matters as Much as How Much

Cash flow isn't just quantities; it's choreography.

A person can have a positive monthly margin and still get hit with overdraft fees because their bills cluster before payday. Or because an annual insurance premium drops like a boulder in a random month. Or because their income is irregular: freelancing, commissions, tips, seasonal work.

Timing problems are stressful because they create the sensation of always being behind, even if the month "works" on paper. The fix is not grit. The fix is to place a small buffer inside the tank—a minimum water level you don't cross.

Think of it as a safety line drawn on the side of the tank: *Below this line, nothing leaves the account except true emergencies.* That one line can prevent a cascade of fees that turn a minor timing issue into a financial infection.

A practical starting point is a buffer equal to one week of burn rate. If your required bills are $2,400 a month, one week is roughly $600. Keep $600 as the minimum. The exact number can change later, but the concept is what matters: your checking account isn't a place to keep "whatever." It's an operating tank that needs pressure.

The ``Invisible'' Expense: Taxes and Withholding Surprises

For W-2 employees, taxes feel automatic because they're withheld. For anyone with side income, freelance work, gig driving, or meaningful investment income, taxes can become the sneakiest drain of all—because the bill arrives *later*.

Cash flow discipline means treating taxes like a bill you pay as you go, not a surprise you wrestle at filing time. The emotional damage of a big April bill is real: it can wipe out months of progress, trigger credit card balances, and create that sinking feeling of "I can't get ahead."

A simple habit prevents it: every time untaxed income hits, immediately siphon a percentage into a separate "tax" bucket (even if that bucket is just a labeled savings sub-account). It's not fun money. It's not investment money. It's money that already has a job.

How to Find Your Leaks Without Becoming a Spreadsheet Person

Some people love tracking. Others would rather floss with barbed wire. The good news: you don't need obsessive detail. You need a few high-signal questions.

Try this for one month:

1. List every fixed bill and its due date.
2. For variable spending, don't categorize everything. Just tag purchases with one of three labels: *Fuel* (keeps life running), *Fun* (worth it), *Fog* (you don't remember why you bought it).

"Fog" is where margin gets murdered. It's not the concert you planned for. It's the half-interested scrolling purchase, the second delivery fee of the week, the "free trial" that quietly matured into a monthly charge you never notice because it's only $12.99.

Fog isn't evil. It's unmanaged.

Once you spot fog, you can narrow the drain without touching the spending that actually makes you happy. That's the point: protect what you *love*, eliminate what you *won't miss*.

Turning the Gap Into a Physical Reality

The most important moment in cash flow is not the end of the month. It's payday.

If you wait to see what's left, you're fighting a psychologically unfair battle: every dollar in your checking account feels available. Humans are excellent at spending what looks spendable. That's not stupidity; it's wiring.

To make the gap real, you need to capture it early, the way a city captures rainwater before it floods the streets.

On payday, the order of operations matters:

1. Money enters the tank.
2. You reserve what must leave soon (bills).
3. You capture overflow (savings) before it has a chance to "become" spending.
4. The rest is for life.

Even if you start small—$25 a week—you're training the system to overflow. You're proving to your nervous system that progress is possible without drama. And you're building the habit that makes later goals easier: you don't save when you feel rich; you save because your system does it by default.

Actionable Takeaway: Build Your One-Page Cash Flow Scoreboard

Tonight, make a one-page snapshot. No apps required.

1. Write your monthly take-home income at the top.
2. Under it, list your fixed expenses and total them. That total is your burn rate baseline.
3. Estimate your variable spending last month (a rough number is fine).
4. Compute your gap:

$$\text{Gap} = \text{Income} - (\text{Fixed} + \text{Variable})$$

5. Convert it to a savings rate:

$$\text{Savings Rate} = \frac{\text{Gap}}{\text{Income}} \times 100\%$$

If the gap is negative, choose one drain to narrow by a specific amount this month (even $50) and write the exact mechanism you'll use (cancel, negotiate, downgrade, cook twice, change a habit trigger). If the gap is positive, pick a number—even a small one—to capture on payday automatically so the overflow becomes a fact, not a hope.

The goal isn't to become a different person. It's to build a system where your money finally does what you've been trying to make it do all along: stay long enough to matter.

2.2 Designing a Spending Plan That Works

Most budgets fail for the same reason most diets fail: they treat you like a robot who can be reprogrammed with shame.

Monday arrives, you open a spreadsheet, and you swear you're done with "wasting money." No more dinners out. No more little treats. No more impulse buys. You'll live on rice and discipline. And for a few days it even feels good—the clean, sharp feeling of being in control.

Then real life shows up.

A friend invites you to a birthday dinner. Your boss drops a late request and cooking becomes a joke. You have a stressful week and your brain starts hunting for relief the way it always does. Suddenly you're not "breaking a budget." You're breaking a promise to yourself. That's where the spiral begins: guilt, then avoidance, then the budget becomes a document you don't open.

A spending plan that works isn't a punishment. It's a design that makes room for being human.

The Mindset Shift: From ``Restrictions" to ``Trade-offs You Actually Choose"

A restrictive budget asks, *What can I cut?* A values-based spending plan asks, *What do I want my money to do for me?*

That sounds soft until you realize it's brutally practical. Your money already has priorities. If you don't choose them deliberately, they get chosen by defaults: convenience, algorithms, advertising, and whatever you're stressed about that day.

The goal is not to spend as little as possible. The goal is to spend *on purpose*—and to stop accidentally spending your future on stuff your present self barely remembers.

A good spending plan does three things at once:

- It keeps the bills paid (because chaos is expensive).
- It captures your gap (because that's how wealth gets built).
- It funds a life you actually like (because otherwise you'll rebel).

If any one of those is missing, the plan collapses. Bills without joy becomes resentment. Joy without a gap becomes debt. A gap without bill stability becomes overdraft roulette.

Why ``I Just Need More Willpower" Is a Trap

Willpower is useful, but it's not a renewable energy source. It's more like phone battery: you start the day with some, and then work, stress, hunger, and decision fatigue drain it.

A spending plan that depends on you making twenty perfect choices a day is a plan that will eventually lose to a Tuesday.

So the real question is: *How do you build a plan that still works when you're tired, busy, and feeling a little impulsive?*

Answer: fewer decisions, clearer rules, and categories that match how you naturally think. Not how a financial guru thinks. Not how a spreadsheet thinks. How *you* think when it's 9:17 p.m. and you're one click away from ordering something you don't need.

The 50/30/20 Rule: Not a Law, a Flashlight

The 50/30/20 rule is a classic guideline:

- 50% to Needs (the bills you can't ignore)

- 30% to Wants (the fun, the comforts, the lifestyle choices)
- 20% to Savings (and often debt payoff)

Used the wrong way, it becomes another stick to beat yourself with. Used the right way, it's a diagnostic tool. A flashlight in a messy room.

If you're far from the numbers, that doesn't mean you're doing life wrong. It means you have constraints: high rent, medical expenses, childcare, student loans, a temporary income dip. The value of the rule is not obedience. It's clarity.

Think of it like stepping on a scale. The scale doesn't judge you. It gives you data. Your plan starts with data.

If your Needs are 70% right now, the plan isn't "try harder." The plan might be "shrink a major drain over the next 12 months" or "increase the faucet" or "temporarily reduce Wants while you build a buffer." The numbers point to the lever.

A Spending Plan Is a Map, Not Handcuffs

A budget is often treated like a fence: cross the line and you've failed. A spending plan works better as a map: it shows you where you are and helps you choose where to go next.

Maps are flexible. If there's traffic, you reroute. You don't slam the steering wheel and decide you're "bad at driving."

Your money needs the same attitude. The plan should handle:

- irregular expenses (car repairs, gifts, annual fees),

- social life (because isolation is a hidden cost),
- higher-spending months (holidays, weddings, travel),
- and the fact that you will sometimes spend for comfort.

A working plan doesn't pretend you won't. It accounts for it and contains it.

The Secret Ingredient: ``Guilt-Free Spending'' as a Category

People don't blow budgets because they don't understand math. They blow budgets because they didn't give themselves permission to be a person.

Guilt-free spending is money you can spend *without doing mental math, without bargaining, without promising you'll be better next week*. It's pre-approved.

This category is not financial irresponsibility. It's pressure relief. Without it, every purchase becomes a moral debate, and debates are exhausting. Exhaustion leads to rebellion. Rebellion leads to "whatever, I'll fix it later."

The plan needs a release valve.

The trick is to put the release valve in the *plan*, not in the moment.

Case Study: The ``Always Broke'' Couple Who Wasn't Actually Broke

Janelle and Marco both worked full-time. Combined take-home pay: about $6,200 a month. On paper, they should have been fine. In reality, they felt like money was constantly slipping away, and they kept floating expenses on a credit card even though their income was solid.

When they finally looked closely, the issue wasn't that they were lavish. It was that their spending had no *shape.*

They had three patterns:

1. *The convenience tax:* takeout on late work nights, grocery delivery fees, rideshares instead of transit "just this once."
2. *The social blur:* a handful of dinners and events each month, each one reasonable, but collectively enormous.
3. *The silent annuals:* car registration, holiday gifts, one annual subscription renewal, and a couple of medical bills that hit like jump scares.

Their solution wasn't extreme cutting. It was category design.

They built a plan with four buckets:

- *Must-Pay Bills* (housing, utilities, insurance, minimum debt payments)
- *Food at Home* (groceries only)
- *Life and Fun* (restaurants, dates, entertainment, social events)
- *Future* (emergency fund and extra debt payments)

Notice what they didn't do: they didn't create twelve fussy categories that required a weekly accounting ritual. They designed categories around their actual friction points.

Then they added one more: *Irregulars*. A monthly amount that existed solely to absorb the "not every month" expenses. Once Irregulars was funded, those surprise bills stopped feeling like personal failure.

Within two months, their credit card balance stopped creeping up. Not because they became more disciplined people, but because the plan finally matched reality.

How to Build Categories That Fit Your Brain

The best spending plan is the one you'll keep using. That means categories should be:

- *Few enough* that you can manage them,
- *Clear enough* that you don't argue with yourself about where something belongs,
- *Honest enough* to capture the real drains.

Start with the big buckets and only split when a bucket becomes a problem.

Here are examples of categories that tend to work for real humans:

- Housing
- Transportation
- Food at home
- Eating out / convenience food
- Health (meds, copays)
- Minimum debt payments

- Subscriptions
- Fun / Social
- Irregulars (gifts, repairs, annual fees)
- Savings / Investing (your "Future" bucket)

If you're thinking, *That looks like a lot,* it can be fewer. The point is not the exact list. The point is to separate the areas where you tend to leak.

Some people leak through food delivery. Others leak through Amazon and hobby spending. Others leak through "small" subscriptions that multiply into a second utility bill. Your categories should shine a light on *your* fog.

Needs vs Wants: A Useful Distinction, Not a Courtroom

"Needs" and "wants" can turn into a fight, especially if you grew up hearing that wanting things makes you irresponsible.

Here's a calmer way to define them:

- A *need* is an expense with serious consequences if you stop paying it.
- A *want* is something you could change, downgrade, or pause without immediate danger.

The line is not always clean. Internet can be both. A car can be both. Childcare is often non-negotiable. Mental health supports can be essential.

So don't treat Needs vs Wants as a moral label. Treat it as a planning tool: if you need to create margin quickly, Wants are usually the easiest drain to narrow *without risking your stability*. If you want to build long-term breathing room, shrinking certain Needs (like housing or transportation) can change your life, but it often requires a bigger move.

The ``Irregulars" Category: Where Plans Go to Die (Unless You Fund It)

Irregular expenses are predictable in the way winter is predictable. You don't know which day it will snow, but you know snow is coming.

A spending plan breaks when it assumes every month is a clean rectangle. Real life is lumpy.

Common irregulars:

- car repairs and maintenance
- gifts and holidays
- travel (even modest)
- annual subscriptions and membership renewals
- medical and dental bills
- work expenses that get reimbursed later

If you don't plan for these, they land on a credit card, and then the card becomes a permanent second budget with interest attached.

So pick a monthly amount—even $50 to start—and call it Irregulars. It's not exciting, which is exactly why it works. It turns financial jump scares into scheduled weather.

The ``No-Overdraft" Rule: Build a Plan That Respects Timing

Remember the tank: even if your monthly math is positive, timing can still wreck you.

A practical spending plan includes a simple operating rule: keep a minimum cushion in checking. That cushion is not savings; it's a timing buffer that prevents fees and stress.

Two additional moves make timing smoother:

- *Due-date awareness:* put every bill's due date on one list, then notice clustering.
- *Cash cushion around autopays:* if multiple bills hit within a few days, treat those days like a storm window and keep extra water in the tank.

This is where a spending plan stops being theory and becomes operations.

Spending ``Rules" That Don't Feel Like Rules

A lot of spending plans die because they try to micromanage behavior. A better approach is to create a few simple defaults that make overspending harder without constant self-denial.

Examples:

- *The 24-hour rule* for non-urgent purchases over a set amount (say $50 or $100).
- *One convenience meal limit* per week (or per workday), with a pre-decided substitute (a freezer meal, a simple grocery list, a go-to cheap takeout).

- *Subscription amnesty day* once a month: cancel or downgrade anything you haven't enjoyed in the last 30 days

These aren't about being strict. They're about preventing "Fog" spending from turning into a second rent payment.

How to Know If Your Plan Works (Without Waiting for the End of the Month)

A working plan gives you feedback early enough to adjust.

Pick one mid-month check-in date—for many people, around day 15 works. Look at three numbers:

- How much you've spent in your most leak-prone category (often food out, shopping, or fun).
- How much is left in checking *relative to upcoming bills.*
- Whether your "Future" bucket received money (because if it didn't, the plan is drifting).

If things are off, the correction is small and immediate: one quieter weekend, one grocery-focused week, pausing a category temporarily. This is how you prevent the end-of-month panic where the only solutions feel extreme.

Actionable Takeaway: Write a One-Page Spending Plan You'll Actually Follow

Do this in one sitting. Set a timer for 30 minutes.

1. Write down your monthly take-home income.

2. Choose 5–8 categories that match your real life. Include *Irregulars* and *Future*.
3. Fill in your fixed bills first (rent, insurance, minimum debt payments). These are your non-negotiables.
4. Decide on a guilt-free spending amount for the month. Put it inside a category like *Fun/Social* so it's planned, not apologized for.
5. Choose a starting "Future" number you can hit even on a messy month. If you're not sure, pick $25 per paycheck. Small is fine; consistent is the win.
6. Add a simple rule for your biggest leak (like a 24-hour rule or a convenience-meal limit).

Then make one promise that's easy to keep: you will check the plan once mid-month and once before the month ends. Not to scold yourself. To steer.

A spending plan that works doesn't make you feel trapped. It makes you feel *safe*. Bills get handled. Fun is allowed. Progress happens in the background. And when temptation shows up—as it always does—you're not negotiating with your future in the heat of the moment. You already decided what matters.

2.3 Organizing Your Financial Operations

If money management feels like a constant test of self-control, there's a decent chance you're not dealing with a character flaw. You're dealing with a bad floor plan.

Most people run their entire financial life through one checking account: paychecks land, bills leave, impulse spending happens, savings is whatever survives. That setup asks one small account balance to do three jobs at once: *pay*, *spend*, and *store*. It's like using your kitchen counter as a fridge, a pantry, and a trash can. You can technically do it, but you'll spend your whole life cleaning up messes.

Wealth-building becomes dramatically easier when you separate *spending money* from *savings capital* in a way your future self can't accidentally sabotage at 11:48 p.m. with a "Why not?" purchase.

The Core Idea: Make the Right Thing the Easy Thing

Earlier, you built the gap: the difference between what comes in and what goes out. The next challenge is keeping that gap from quietly sliding back into your everyday spending.

This is where operations matter. Operations isn't glamorous, but it's powerful: the right accounts, in the right roles, with money flowing between them on purpose.

A well-organized banking setup does something subtle and priceless: it turns "saving" from an emotion into a *default*. You stop negotiating with yourself every payday. The system makes fewer things possible—and that's exactly why it works.

The Hub-and-Spoke Architecture

Picture your money like air traffic. If every plane tries to land wherever it wants, you get chaos. A hub airport creates order: arrivals come in, then get routed to the right destinations.

Your financial hub is a central checking account—a purpose-built operations center where income lands and bills get paid.

From there, money routes outward to "spokes" that each have a single job.

The simplest effective version has three spokes:

- *Bill Pay* (a buffer for fixed expenses and due dates)
- *Spending Cash* (your guilt-free, day-to-day money)
- *HYSA Vault* (high-yield savings account for emergency fund and near-term goals)

The point isn't to collect accounts like trophies. The point is to create *containers* that prevent category confusion.

Account Types: What Each One Is Actually For

A quick, practical definition in plain language:

- *Checking account:* an operations account. It's built for transactions: deposits, bill pay, debit card swipes, transfers.
- *Savings account (especially a HYSA):* a storage account. It's built for holding money safely while earning interest, with fewer temptations and fewer daily transactions.

If your savings sits in the same checking account as your weekend plans, it's not really savings. It's *available spending that hasn't happened yet*.

A HYSA is simply a savings account that tends to pay a higher interest rate than the sleepy savings accounts many banks pair with checking. The interest rate won't make you rich overnight, but it does something psychologically important: it signals that the money is *parked*, not waiting to be spent.

Where the Diagram Fits: The Flow That Makes This Work

The magic is the routing—the *direction* money moves. This is why the arrows matter more than the account labels.

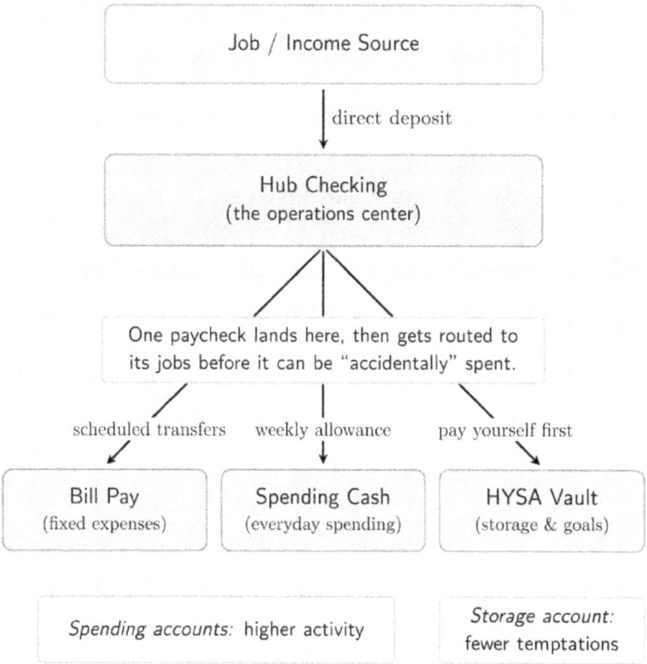

Why This Beats ``Just Be Disciplined"

This structure works because it matches how your brain already behaves.

When money is pooled, every dollar looks identical. Your checking balance becomes a psychological permission slip: "I have $1,800, so I can afford it." But that number may include rent, insurance, and next week's groceries. Pooled money creates fake confidence, followed by real consequences.

Separated money creates clarity. If your Spending Cash account has $120 left until Friday, you don't need a pep talk. You need a plan. The account balance *is* the plan.

A good system does not rely on constant self-control. It relies on *physical separation* and *predictable rules*.

The Bill Pay Spoke: Where Stability Lives

The Bill Pay spoke exists for one reason: to make sure your essential obligations don't compete with your daily spending.

Many people pay bills directly out of the same account they use for coffee, apps, and random store runs. That's a timing game you eventually lose. A surprise autopay can turn a normal week into overdraft fees and stress-texting your bank.

The Bill Pay account (or sub-account) is where you park the month's fixed expenses. When it's funded, your nervous system relaxes. You stop doing that constant mental math where you try to remember what's about to hit.

A clean way to run it:

- Total up your fixed expenses for the month.
- Divide by the number of paychecks you get.
- Transfer that amount into Bill Pay each payday.

It's boring. It's also how you stop living at the mercy of due dates.

The Spending Cash Spoke: Freedom With Guardrails

Spending Cash is the account that prevents the plan from feeling like a cage.

This is where you put the money you're allowed to spend freely: restaurants, fun, small purchases, whatever makes life enjoyable. When you spend from this account, you don't have to ask permission. The permission already happened when you funded it.

The guardrail is simple: when it's empty, you stop. Not because you're being punished, but because you're protecting rent, keeping your gap, and making sure future-you doesn't get stuck cleaning up the mess.

If you've ever tried to budget with one account, you know the feeling: you spend money, then you feel guilty because you can't tell whether you just spent grocery money or savings money or "we'll see" money. Spending Cash removes that ambiguity.

The HYSA Vault: Where Money Becomes Capital

The HYSA Vault is the account that changes your identity from "someone who earns money" to "someone who accumulates capital."

Capital is money that has been captured and set aside to do future work: absorb emergencies, pay for planned expenses, reduce reliance on debt, and eventually fund investing.

The Vault has two crucial features:

- *It's not in your face.* You shouldn't see it every time you open your banking app to check whether you can afford sushi.
- *It's liquid.* You can access it when real life happens, without selling investments at the wrong time.

A HYSA is also a psychological moat. Not an impenetrable one, but enough friction that you pause before treating savings like spending.

A Concrete Example: How This Fixes ``I Save, Then I Unsave"

Nia gets paid biweekly. She's done the responsible thing a dozen times: she moves $200 into savings on payday, feels proud, and then two weeks later she moves $160 back because her checking ran low. She tells herself she's bad at saving, when the truth is simpler: her money has no assigned jobs.

Here's what changes with Hub-and-Spoke:

1. Payday hits Hub Checking.
2. Immediately, $X goes to Bill Pay (so rent and insurance don't sneak-attack her checking balance).
3. $Y goes to Spending Cash (so she can live without constant anxiety).
4. $200 goes to the HYSA Vault (the gap becomes real).

The difference is that the HYSA transfer no longer competes with bills. Bills already have a home. Spending already has a home. So savings stops being the flexible, sacrificial category that gets raided whenever anything is tight.

When people say, "I can't save," they often mean, "My bills and my spending are fighting in the same room."

Choosing Banks and Accounts: What Actually Matters

This isn't about picking a fancy institution. It's about choosing features that reduce friction and fees.

Prioritize:

- *Low or no monthly fees* (fees are negative interest).
- *Overdraft policy you understand* (and ideally controls you can set).
- *Alerts and notifications* for low balances and large transactions.
- *Online bill pay* and easy scheduled transfers.
- *ATM access* that fits your life if you use cash.

And for peace of mind, know the basic safety net: in the U.S., deposits at FDIC-insured banks and federally insured credit unions (NCUA) are generally protected up to limits by ownership category. That doesn't mean you should ignore bank choice, but it does mean your savings strategy doesn't require hiding cash in a drawer to feel safe.

A Simple ``Bucket" Table You Can Copy

If you like a concrete template, this is a clean starting layout. Adjust the amounts, keep the roles.

Account	*Job Description*
Hub Checking	Paycheck lands here. Routes money outward. Not a storage unit.
Bill Pay	Holds fixed expenses and recurring obligations. Built for stability and timing.
Spending Cash	Your allowed, guilt-free spending. When it's empty, you pause.
HYSA Vault	Emergency fund and near-term goals. Money you're protecting from casual spending.

The Most Common Failure Mode (and How to Avoid It)

The failure mode isn't setting up the accounts. It's leaving the Hub Checking balance so high that you treat it like a second Spending Cash account.

If your hub always has "extra" money, you'll spend it. Not because you're irresponsible, but because you'll interpret it as available. The hub should feel like an airport terminal: money arrives, waits briefly, then departs to its destinations.

A useful rule of thumb: keep the hub lean. Enough buffer to avoid overdrafts and handle timing, but not so much that it becomes a temptation pile.

Actionable Takeaway: Set Up Your Banking Floor Plan in One Hour

Block one hour this week and do the operational work. You don't need perfection; you need separation.

1. Choose your Hub Checking (existing is fine if it's fee-friendly).
2. Create (or repurpose) three destinations: Bill Pay, Spending Cash, and a HYSA Vault.
3. Decide on two numbers: the per-paycheck transfer to Bill Pay (fixed expenses divided by paychecks) and the per-paycheck transfer to HYSA (your gap capture, even if it's small).
4. Pick a Spending Cash amount that allows you to live without resentment.
5. Turn on alerts for low balances and large transactions.

When your money is physically separated into roles, you stop needing constant motivation. You can have a rough week, a busy month, even a few impulsive moments—and still move forward, because the system quietly keeps doing its job.

2.4 Putting Your Money on Autopilot

There's a version of personal finance advice that sounds noble and quietly ruins people: *"Just be disciplined."*

Discipline is great. It's also unreliable. Not because you're weak, but because you're human. You have long workdays, sick days, messy emotions, broken phones, surprise deadlines, and the kind of fatigue that makes "I'll deal with it later" feel like a rational financial strategy.

Automation is what happens when you stop asking your tired brain to be the hero.

A well-built autopilot system takes your best intentions and locks them in while you're calm, rested, and thinking clearly. Then, when life gets loud, the system keeps moving money in the right direction without requiring a single pep talk.

The Real Enemy: Decision Friction

Most people don't fail to save because they don't care. They fail because saving requires a chain of decisions at exactly the wrong moments.

Consider the manual approach:

- You get paid.
- You remember you meant to save.
- You log into your bank.
- You check your balance.
- You start doing mental math about upcoming bills.
- You decide to transfer "whatever is safe."
- You feel uncertain, so you transfer less.
- You promise you'll make up for it next paycheck.

This isn't laziness. It's friction. Saving is competing with everything else that needs your attention. The friction wins.

Automation flips the script. It makes saving the path of least resistance and makes overspending require extra steps.

``Pay Yourself First" Means First in Time, Not First in Theory

Pay Yourself First is a simple concept with a strict definition: savings happens *before* spending has a chance to claim the money.

Lots of people say they pay themselves first and then save what's left. That's not first. That's last. "Whatever's left" is a category that gets eaten alive by a thousand reasonable expenses.

First means time order:

1. Paycheck arrives.
2. Savings and investing leave automatically.
3. Bills get paid on schedule.
4. The remaining money is yours to spend.

The beauty is psychological. You stop feeling like saving is what you do only when you're being "good." It becomes a normal bill you pay to your future.

The Autopilot Ladder: What to Automate (and in What Order)

Not everything should be automated immediately. The goal is stability first, then growth.

A clean sequence looks like this:

- *Tier 1: Bills that protect your life.* Rent/mortgage, utilities, insurance, minimum debt payments. Automation here prevents late fees and chaos.
- *Tier 2: Your buffer.* A small transfer to savings each payday until you have a cushion.

- *Tier 3: Investing and extra debt payoff.* Once cash flow timing is stable, you automate the wealth-building moves.

This order matters because automation is powerful, but it's not magical. If you automate aggressively without a buffer, you can create overdrafts. Then you'll blame automation for what was really a missing safety margin.

Direct Deposit Splits: The Cleanest Autopilot Hack

If your employer allows split direct deposit, you've been handed one of the most effective wealth-building tools that doesn't look like a wealth-building tool.

Instead of depositing your full paycheck into Hub Checking and hoping you transfer savings later, you route money *before you touch it*:

- A fixed amount (or percentage) to your HYSA Vault.
- The rest to Hub Checking for bills and spending.

Why this works: it bypasses the most dangerous moment in personal finance—the moment you see a large balance and your brain starts assigning it a lifestyle.

Split deposit turns savings into a default, not a decision.

If you don't have split deposit, you can simulate it with an automatic transfer scheduled for payday (or the day after). The split is cleaner, but both work.

Autopay Is a Knife: Useful, Sharp, and Not for Waving Around

Autopay can save you money and stress by preventing late fees and missed payments. It can also hurt you if it pulls money when your checking balance is low, triggering overdraft fees or returned payment fees.

The difference between autopay that helps and autopay that harms is a simple design choice: *do you have a cash cushion and a timing plan?*

A smart autopay setup looks like this:

- Fixed bills are paid automatically from a Bill Pay buffer or a well-funded hub.
- Payment dates are aligned with your pay cycle when possible.
- Alerts are turned on so you know before a payment hits.

Autopay is at its best when it's boring: the payment happens, you barely notice, your credit stays clean, your brain stays quiet.

Autopay is at its worst when it becomes a monthly jump scare.

Case Study: The Teacher Who Saved More Without Feeling It

Alana is a public school teacher. She didn't have a spending problem in the dramatic sense. No luxury shopping sprees. No expensive hobbies. Still, she ended most months with a small credit card balance and a familiar feeling: *I make decent money—why does it feel like I'm barely treading water?*

Her issue was timing and effort. She intended to save, but she was always tired at the exact moment she needed to make the decision. Paydays were followed by catch-up spending: groceries, gas, small classroom supplies, a dinner out because it had been a week.

She built an autopilot system in one afternoon:

1. She set up split direct deposit: $150 per paycheck to her HYSA Vault, the rest to Hub Checking.
2. She scheduled a transfer from Hub Checking to Bill Pay the day after payday to cover fixed expenses.
3. She put her credit card on autopay for the *statement balance* but only after building a small checking cushion.
4. She turned on low-balance alerts for Hub Checking and for the credit card.

In month one, nothing dramatic happened. That was the point. But something subtle changed: her savings grew without requiring motivation. She didn't feel like she was "trying." She felt like the system was quietly doing what she'd been trying to force.

By month three, she had $900 in the HYSA that hadn't been there before. Not because she discovered a new level of discipline, but because she removed the decision points that used to fail.

The ``Default Effect": Why Autopilot Is So Powerful

People like to think they make fresh decisions all the time. In reality, defaults shape behavior more than motivation does.

When a savings transfer is automatic, the default becomes *saving*. When saving is manual, the default becomes *spending now and hoping later works out*.

Defaults work because they harness inertia. Inertia is usually blamed for procrastination, but it's morally neutral. It simply means things tend to keep going the way they're set up. If your system is pointed toward wealth-building, inertia becomes your ally.

This is one reason automatic enrollment in retirement plans has been so effective: when participation is the default, more people participate. When opting in requires action, many never get around to it.

So the goal is not to become a person who overcomes inertia through constant effort. The goal is to set the direction of inertia correctly.

The ``Save More Tomorrow" Idea: Raise Your Savings Without Feeling Poor

One of the smartest psychological hacks in personal finance is to increase savings when your income increases.

If you get a raise and immediately upgrade your lifestyle, you stay on the same treadmill with nicer shoes. If you route part of the raise into savings automatically, you get wealthier without feeling deprived, because you never got used to spending the full raise.

A practical way to do this:

- Every time your paycheck increases, increase your automatic savings by 30–50% of the net raise.

Example: you get an extra $200 a month after taxes. Increase your automatic savings by $80 or $100. You still feel the raise—your spending money increases—but your future also gets paid.

This works because your brain adapts quickly to new normal. If the savings happens before you adapt, you don't miss it.

Automation Without Chaos: The Cushion Rule

The fastest way to hate automation is to set it up without protecting your checking account.

A simple rule keeps you safe: keep a minimum operating cushion in Hub Checking. This is not your emergency fund; it's your turbulence buffer. It prevents autopays and transfers from causing overdrafts during weird timing weeks.

A starting cushion could be:

- one week of fixed expenses, or
- a flat number like $300–$1,000 depending on your burn rate and variability.

If your cash flow is tight, start with a smaller cushion and build it gradually. The point is to avoid the situation where one automated payment triggers a chain reaction of fees that punishes you for trying to be responsible.

Automation should make your life smoother. If it's making your life spikier, it's not a character issue; it's a settings issue.

What to Automate If Your Income Is Irregular

If your income varies (freelance, commission, tips), traditional "same amount every payday" automation can feel risky.

You can still automate; you just automate differently:

- *Automate percentages, not fixed dollars* where possible (some payroll systems and some banks allow this).
- *Use a weekly transfer* to savings that's intentionally modest, then make an extra transfer on good weeks.
- *Automate bills from a buffer account* funded during higher-income periods.

The goal is to avoid overcommitting on a low-income month while still keeping the habit alive. Consistency matters more than intensity.

The Autopilot Checklist: Your Money Routes Like a Conveyor Belt

A clean autopilot setup often has a simple rhythm:

- Payday: paycheck arrives (split deposit if possible).
- Day after payday: automatic transfers route money into Bill Pay and HYSA.
- Mid-cycle: credit card autopay happens.
- Throughout the month: alerts notify you if balances dip below your cushion.

Once this is running, you stop doing daily financial triage. You don't have to constantly remind yourself to be responsible. The system becomes the responsible friend who shows up every time.

Actionable Takeaway: Set Up a ``Two-Transfer Autopilot" This Week

You don't need a dozen automations to change your life. Start with two.

1. *Transfer #1: Pay Yourself First.* Set an automatic transfer (or split direct deposit) to your HYSA Vault on payday (or the day after). Choose an amount that is small enough to be sustainable in a messy month. If you're unsure, start with $25 per paycheck.
2. *Transfer #2: Stabilize Bills.* Set an automatic transfer from Hub Checking to your Bill Pay buffer each payday, based on your fixed expenses divided by paychecks.

Then turn on two alerts:

- low-balance alert on Hub Checking (to protect against overdrafts),
- payment reminder alert for any large autopay.

If you do nothing else for the next 30 days, let those two transfers run. Watch what happens to your stress level when saving and bills stop depending on your mood. The goal isn't to become perfectly disciplined. It's to build a system that behaves well even when you don't feel like it.

2.5 The Sleep-Well-At-Night Fund

There's a moment that changes how you move through the world: your car makes a sound it shouldn't, your stomach drops, and then you remember you have cash set aside. The sound is still annoying, but it's not a crisis. You don't start doing mental gymnastics about credit cards. You don't start bargaining with your future. You just handle it.

That's what an emergency fund really buys. Not a return. Not a flex. *Relief.*

People love to argue about investing versus saving, as if these are rival sports teams. The emergency fund isn't on the field. It's the stadium lighting. When it's working, you barely notice it. When it fails, everything else becomes harder and riskier.

The Emergency Fund Is Insurance, Not an Investment

Insurance is money you pay to make sure a rare-but-painful event doesn't break you. You don't complain that homeowners insurance has a "low yield." You're grateful you don't need it most years.

Your emergency fund is the same category of tool. It sits in safe, liquid cash (often a high-yield savings account), ready to absorb shocks that would otherwise ricochet through your life: late fees, overdrafts, credit card balances, missed rent, selling investments at the wrong time, or taking a desperate loan with ugly terms.

The fund's job is to keep small problems small.

And it's not only about emergencies in the dramatic sense. It's about the everyday financial shocks that show up with a straight face and demand cash: a dental bill, a broken phone, a plane ticket for a funeral, a medical deductible, a job gap, a surprise tax bill, a landlord deciding to raise rent.

Life doesn't ask permission.

Why Smart People Still Get Wrecked by Small Shocks

It's tempting to think money stress is always about income. Sometimes it is. But a surprising amount of financial pain is caused by *fragility*—the lack of a buffer.

Without a buffer, you pay "penalty prices" for normal problems:

- a late fee because the timing was off,
- an overdraft fee because autopay hit before payday,
- credit card interest because you had to float a repair,
- a missed appointment fee because you couldn't take time off work,
- a high-interest loan because you needed cash today, not next month.

These penalty prices create a brutal feedback loop. You spend money because you're broke, and you stay broke because you're spending money on the consequences of being broke.

A proper emergency fund breaks that loop. It stops your finances from behaving like dry tinder—one spark, and everything catches.

The $400 Problem (and Why It's a Useful Benchmark)

A lot of financial advice is so aspirational it becomes alien. "Save six months of expenses" is solid guidance, but it can feel like being told to climb a mountain when you're still tying your shoes.

So here's a more grounded starting question: *Could you cover a $400 surprise expense without spiraling?*

That number matters because it's small enough to be realistic and common enough to be relevant. It's a tire. It's a car battery. It's a copay plus a prescription. It's the kind of thing that happens in normal life, not worst-case life.

If $400 would push you into debt or panic, the emergency fund doesn't need to be a grand project. It needs to be a fast win:

- $400 is your first milestone.
- $1,000 is your next.
- Then you move toward months of expenses.

Those early milestones are not random. They buy you stability quickly. They also build trust with yourself: you start to believe your system works.

How Much Do You Actually Need? Start With Burn Rate

An emergency fund is sized based on your *burn rate*—the amount your life requires each month to keep running.

Burn rate is not your ideal budget. It's your real baseline: housing, utilities, insurance, minimum debt payments, basic food, transportation to earn income, and anything else that would cause immediate trouble if you stopped paying.

Once you have that number, sizing becomes less mystical. You're no longer guessing. You're buying a specific amount of time.

$$\text{Emergency Fund Target} = \text{Monthly Burn Rate} \times \text{Months of Coverage}$$

The remaining question is how many months makes sense for *your* life. That depends less on how brave you are and more on how predictable your income is.

A Practical Sizing Guide (Without Making It Complicated)

Instead of pretending there's one perfect number, treat emergency fund sizing like choosing the right coat. The weather matters.

Here are realistic guidelines:

- *3 months* if your income is stable, you have few dependents, your job market is healthy, and your fixed expenses are manageable.
- *6 months* (*or more*) if your income is variable, you're self-employed, you're in a commission-heavy role, you have dependents, you're supporting family, you have a medical situation, or your job would be hard to replace quickly.

There's no shame in starting at one month and building from there. The fund is a ladder, not a leap.

One important nuance: if you have high-interest debt, you may be tempted to throw every spare dollar at it. Debt payoff is important, but paying off debt *without* a buffer often backfires. You'll celebrate a lower balance and then put the next emergency right back on the card.

A small emergency fund is what keeps debt payoff from turning into a frustrating game of financial whack-a-mole.

Case Study: The Layoff That Didn't Become a Disaster

Devon worked in tech operations. Good income, decent benefits, and the quiet belief that layoffs were something that happened to other people.

Then it happened to him on a Tuesday morning. A calendar invite with a vague title. A short meeting. Access shut off by lunch.

Devon wasn't reckless, but he was optimized for normal times: he kept about $800 in his checking account and invested everything else he could. His investing account looked respectable. His cash reserves looked like a rounding error.

He did one thing right six months earlier: he had built a *Sleep-Well-At-Night* fund of about three months of burn rate in a high-yield savings account. He almost felt silly about it at the time, like he was being overly cautious.

That fund changed the tone of the entire crisis.

Instead of grabbing the first job offer out of fear, Devon took two weeks to update his resume properly, reach out to contacts, and apply strategically. He kept paying bills without juggling due dates. He didn't sell investments at a bad time. He didn't rack up credit card debt and then have to pay interest as a punishment for being unlucky.

The layoff still hurt. But it stayed a *temporary setback* instead of becoming a financial derailment.

That's the emergency fund's real win: it gives you *time*. Time to make good decisions.

Where to Keep It: Liquid, Boring, and Hard to Accidentally Spend

The emergency fund should be:

- *Liquid:* accessible within a day or two.
- *Safe:* not subject to market swings.
- *Separated:* not sitting in the same place as your everyday spending.

A high-yield savings account usually fits these requirements. It earns some interest, but more importantly, it creates a psychological boundary. It's your Vault, not your wallet.

Avoid keeping your emergency fund in:

- stocks (because markets can drop at the exact moment you need cash),
- retirement accounts (because access can trigger taxes and penalties),
- your main checking account (because it will get quietly absorbed into lifestyle spending).

This is one of those moments where "optimal" is the enemy of "effective." The most effective emergency fund is the one that stays intact until you need it.

FDIC/NCUA Insurance: The ``Is My Cash Safe?" Peace of Mind Layer

A good emergency fund should help you sleep, not give you new worries.

In the U.S., many bank deposits are protected up to certain limits if the institution is insured. Banks and credit unions have different insurance systems:

- FDIC insurance for banks,
- NCUA (share insurance) for federally insured credit unions.

Coverage is generally calculated per depositor, per institution, and by ownership category. The details matter if you have very large balances, but for most people building an emergency fund, the key idea is simple: choose an insured institution, and you're using a structure designed to protect depositors.

If you ever build a very large cash position, you can spread it across institutions or categories to stay within coverage limits. But the biggest immediate win is simply not storing your emergency fund in a place where it can evaporate through spending or market volatility.

What Counts as an ``Emergency'' (So You Don't Eat Your Future)

An emergency fund isn't sacred in the religious sense, but it should be protected with a clear definition. Otherwise, it becomes the adult version of a cookie jar.

A good definition:

> An emergency is an unplanned expense or income disruption that threatens your stability if you don't pay it.

That includes:

- job loss or reduced hours,
- urgent medical costs,
- essential car repairs,
- necessary travel for a family emergency,
- critical home repairs (leaks, heating failures).

It does not include:

- predictable annual expenses (those belong in your Irregulars category),
- vacations and gifts (planned fun is not an emergency),
- upgrades you want but don't need.

If you dip into the fund, the rule is simple: you refill it. Not with shame, with a plan. Emergencies are allowed; chronic emergencies are a sign the fund is undersized or your cash flow is too tight.

How to Build It Without Feeling Like Your Life Is On Pause

The emergency fund can feel emotionally heavy because it's not a shiny goal. You can't wear it. You can't post it. It's the opposite of flashy.

So make it satisfying in a different way: turn it into a scoreboard.

Pick milestones that create momentum:

- $400
- $1,000
- 1 month of burn rate
- 2 months
- 3 months

Then automate the contributions. Your autopilot system is perfect for this: every payday, a transfer to the HYSA Vault. If you get a windfall (tax refund, bonus, gift), decide a percentage to allocate to the fund until it's complete.

This is where the "Save More Tomorrow" idea shines. If your income increases, route part of the increase into the fund. You won't feel deprived because you weren't living on that money before.

The Weird Upside: Emergency Funds Increase Your Risk Tolerance in a Good Way

Here's a counterintuitive truth: holding cash can make you a *better* investor later.

When you have a proper buffer, you're less likely to panic-sell investments during a downturn because you need cash for life. You're also more willing to take calculated career risks—apply for a better job, negotiate harder, move to a new city, start a side project—because you're not one missed paycheck away from catastrophe.

The fund isn't conservative in the timid sense. It's conservative in the strategic sense. It creates a stable base so you can be bolder where it actually pays.

Actionable Takeaway: Build Your Fund in Three Moves

Do these three moves this week:

1. *Calculate your burn rate.* Add up the monthly expenses that would cause immediate trouble if you missed them.
2. *Pick your first target.* If you don't have any buffer, choose $400. If you have that, choose $1,000. If you have that, choose one month of burn rate.
3. *Automate a payday transfer into your HYSA Vault.* Make it small enough to be sustainable and large enough to be real. If you're stuck, start with $25 per paycheck and raise it by $5 each month until it feels meaningful.

Then give the fund a name that makes you respect it. Not "Emergency Savings." Something that matches its job: *Rent Stays Paid, No Panic, Time Buffer, Sleep Well.* When the next surprise arrives—and it will—you'll feel the difference immediately: the problem stays in its lane, and your life keeps moving.

2.6 Defending Against Digital Threats

Automation is a financial superpower. It's also a neon sign.

When your bills pay themselves, your savings fills itself, and your accounts are linked together like a neat set of pipes, you've created something worth stealing. Not because you're rich—because you're *connected*. A scammer doesn't need you to have millions. They need a working login, a distracted moment, and a system that moves money quickly once they're inside.

That's the uncomfortable truth of modern money: personal finance is now partly cybersecurity. The good news is that you don't need to become a tech expert. You need a handful of habits that are boring, repeatable, and brutally effective.

Think of this as digging a moat around the castle you've built.

The Three Ways People Actually Lose Money Online

Fraud stories often sound like freak accidents. In reality, most successful attacks fit into a few patterns:

- *Stolen credentials:* your password gets reused, guessed, leaked in a breach, or tricked out of you.
- *Social engineering:* someone convinces you to do the transfer yourself (because if *you* approve it, the bank's defenses matter less).
- *Account takeover through your email/phone:* if someone controls your email inbox or mobile number, they can often reset everything else.

Notice what's missing: Hollywood hacking scenes. Most fraud is not genius. It's persistence plus psychology.

The goal is not to become unhackable. The goal is to become inconvenient. Criminals prefer easy targets. Your job is to stop being one.

Your Email Account Is the Master Key

If someone gets into your email, they can reset passwords for banks, payment apps, shopping sites, and even your tax accounts. It's the master key under the doormat.

So the first defense step isn't a fancy finance app. It's protecting the email account tied to your finances.

Minimum standard:

- Two-factor authentication (2FA) on your primary email.
- A unique, long password (not used anywhere else).
- Recovery settings you've actually reviewed (backup email, phone number).

If you do only one security upgrade this month, do it here.

Two-Factor Authentication: The Seatbelt You Stop Noticing

Two-factor authentication (2FA) means logging in requires two different kinds of proof:

- something you know (a password),
- plus something you have (a one-time code or a security key),

- or something you are (biometrics).

Why it matters: passwords get stolen. Often in bulk. Often silently. 2FA is what prevents a stolen password from becoming a stolen account.

If your bank, brokerage, email provider, retirement account, and payment apps offer 2FA, turn it on. Treat it like locking your front door.

Which 2FA is best?

Not all 2FA is equal.

- *Best:* security keys (physical devices) when available, especially for high-value accounts.
- *Very good:* authenticator apps (time-based codes).
- *Better than nothing:* SMS text codes.

SMS is common because it's easy, but it can be vulnerable to SIM-swapping and other phone-number takeover scams. If an authenticator app option exists, prefer it. If SMS is the only option, use it rather than leaving 2FA off.

Passwords: Stop Trying to Remember Them

A "strong password" is not one that looks like keyboard soup. It's one that is:

- long,
- unique,
- and not reused.

Reusing passwords is the digital equivalent of using the same key for your house, your car, and your office—and then giving copies to every website you've ever signed up for. When one gets compromised, everything becomes vulnerable.

This is why password managers are such a big deal. They let you use unique, long passwords without relying on memory.

A good rule: use a password manager, generate unique passwords for financial accounts, and protect the password manager itself with a long passphrase and 2FA.

A practical note about changing passwords

You may have heard "change your password every 90 days." Modern guidance is more nuanced. Forced, periodic changes can lead people to make weaker passwords or predictable variations.

A sane rule that fits real life:

- Change passwords immediately if you suspect compromise, receive a breach notice, or see unusual activity.
- Otherwise, focus on unique long passwords and 2FA.

Security is not a calendar ritual. It's a containment strategy.

Phishing: The Scam That Looks Like Customer Service

Phishing is when a scammer pretends to be a real institution—your bank, Amazon, the IRS, your employer—and tries to get you to click, log in, or share sensitive information.

Modern phishing is slick. It can look grammatically perfect. It can copy logos and formatting. It can arrive via email, text, social media DMs, or phone calls.

The best defense is a simple behavioral script:

- *Recognize:* unexpected urgency, threats, "verify now," weird sender addresses, mismatched links, attachments you weren't expecting.
- *Resist:* don't click. Don't reply. Don't call the number in the message.
- *Verify independently:* open your bank app directly, type the real website yourself, or call a known number from the back of your card.

The central trick of phishing is to hijack your attention and shorten your decision window. If you slow down, you win.

Case Study: ``It Looked Like My Bank"

Sam got a text on a Friday afternoon: "Unusual charge detected. Reply YES to confirm." The timing was perfect—he had just used his card at a gas station. He replied YES.

A call came minutes later from "the bank," complete with a phone number that matched the bank's name on caller ID. The person was calm, professional, and helpful. They "verified" Sam's identity and said they needed to send him a one-time code to confirm his account. Sam read the code aloud.

Within ten minutes, his bank login had been changed and a transfer was initiated.

Here's what happened: the scammer was logging into Sam's account and used the one-time code to get past the security checkpoint. Sam thought he was proving his identity to the bank. He was actually handing the bank the keys to the thief.

Sam eventually recovered most of the money, but it was a brutal week: phone calls, forms, frozen cards, missed autopays, anxiety every time his phone buzzed.

The fix wasn't learning obscure tech. The fix was one rule:

> Never share one-time codes with anyone, even if they sound like your bank.

A legitimate institution doesn't need you to read back a code that exists to protect *you from them being impersonated.*

Your Phone Number: Protect It Like a Financial Asset

Your phone number is now part of your identity. If a scammer takes it over, they can intercept SMS 2FA codes and reset passwords.

Steps that make phone takeover harder:

- Set a carrier account PIN (separate from your phone unlock code).
- Ask your carrier about extra protections (port-out protection, account freeze options).
- Use authenticator-app 2FA where possible instead of SMS.

If you've ever had your phone service suddenly stop working, treat it as a potential emergency. Call your carrier using a known number (not one from a text) and verify your account status.

Credit Freezes: A Simple Move That Stops a Whole Category of Damage

Identity theft isn't only about stealing money today. It's also about borrowing money in your name tomorrow.

A credit freeze blocks most new lenders from pulling your credit report, which makes it much harder for someone to open new accounts using your information. It doesn't affect your credit score, and you can lift it temporarily when you need to apply for credit.

A practical approach:

- Freeze your credit at the major credit bureaus.
- Store the PINs or access details in your password manager.

This is one of the highest-impact, lowest-effort defenses available to everyday people.

Autopilot Adds Risk Unless You Add Alerts

Automation moves money fast. That's great when it's your system. It's terrible when it's a thief.

Alerts are how you regain reaction time.

Turn on:

- login alerts (new device logins),

- transaction alerts (any transfer, any debit over $X),
- low-balance alerts (to protect you from fraud-triggered overdrafts),
- credit card alerts for card-not-present purchases.

Set the thresholds low at first. You can always raise them. Early on, you want signal, not silence.

When fraud happens, the difference between catching it today and catching it next month is the difference between a contained mess and a full-blown cleanup.

The ``Recovery Plan" You Want Before You Need It

In a fraud situation, people lose time because they're improvising under stress. A simple recovery plan reduces panic and speeds up action.

Your plan can be one note in your phone or password manager:

- the phone number on the back of your debit/credit cards,
- your bank's fraud hotline,
- the list of accounts that are linked for transfers,
- where you keep your emergency fund.

If identity theft occurs, there are official workflows that guide reporting and recovery. The key point is not memorizing bureaucracy; it's knowing *there is* a structured process so you don't waste time chasing random advice from strangers online.

One more important detail: scammers also impersonate recovery services. If someone contacts you offering to "help you recover your money" for a fee, treat it as suspicious and verify independently.

The Money Hygiene Rules That Actually Matter

If you want a security posture that's strong without being paranoid, focus on these rules:

1. Protect your email with 2FA and a unique password.
2. Use a password manager and unique passwords for financial accounts.
3. Turn on 2FA everywhere, preferring authenticator apps or security keys.
4. Never share one-time codes.
5. Don't click links in urgent financial messages; verify independently.
6. Freeze your credit.
7. Turn on alerts so you find out fast when something moves.

This is not a hobby. It's maintenance, like changing the oil in a car. Neglect is expensive. Habit is cheap.

Actionable Takeaway: A 30-Minute Digital Moat Sprint

Set a timer for 30 minutes and do the highest-leverage actions first:

1. Enable 2FA on your primary email and your main bank account.
2. Install a password manager (if you don't have one) and change the passwords on your bank, email, and payment apps to unique, long passwords.
3. Turn on transaction alerts for your checking account and credit card.

Then, schedule one more block of time this week to freeze your credit and add 2FA everywhere else.

When your finances are automated, your life gets easier. When your security is automated—alerts, 2FA prompts, login notifications—your money stops being a soft target. You can build your cash flow engine, let it run, and still sleep like someone who knows the castle has walls.

Chapter 3

Crushing Debt and Mastering Credit

Most people treat debt like a weather event-something that just happens to them and must be endured. But debt is actually a product you buy, and like any product, it comes with a price tag. In this chapter, we stop looking at monthly payments as 'bills' and start seeing them as the mathematical obstacles they are. We will peel back the layers of the banking industry to show you exactly how lenders profit from your minimum payments, how to hack the credit scoring system to work in your favor, and how to deploy a strategic attack plan that eliminates your balances for good.

3.1 The Math Behind Borrowing

Picture money as a tool you can *rent*. When you borrow, you're not buying money-you're leasing it, and the lease has a price tag. That price tag is interest, and it's one of the most efficient wealth-repellent forces ever invented because it's quiet. It doesn't kick down your door. It just skims a little off the top of your future.

The trap isn't that borrowing exists. The trap is that the math is designed to feel harmless at human scale while it compounds like a lab experiment.

APR: the ``rental fee" that hides in plain sight

APR (annual percentage rate) is best understood as the sticker price of borrowing expressed as a yearly percentage. If a loan is a car, APR is the "cost to drive it off the lot and keep it running," not just the engine's raw horsepower. On credit cards it maps closely to the interest rate you feel month to month; on some loans it can also reflect certain fees spread over the year, which is why APR is the number regulators require lenders to disclose: it's supposed to make different products comparable.

The part most people miss is that APR isn't charged annually in one dramatic swoop. It's typically applied in smaller time slices-daily on credit cards, monthly on many installment loans-so your balance becomes a living thing. You're not paying a bill; you're managing a creature that gains weight if you feed it and also gains weight if you ignore it.

A rough translation from APR to monthly rate is:

$$r \approx \frac{\text{APR}}{12}$$

So a 24% APR credit card is roughly 2% per month. That doesn't sound like much until you remember what it's applied to: the unpaid balance, every month, like a meter running in the background.

If you borrow $5,000 at about 2% per month, the first month's interest is roughly:

$$\$5{,}000 \times 0.02 = \$100$$

That $100 is the entry fee to even *begin* paying down what you owe. If your payment is $150, only $50 is doing real work. If your payment is $110, you're crawling. If your payment is $100, you're on a treadmill that never moves.

Amortization: why the early payments feel like a scam (even when it's ``normal")

Ever look at a mortgage statement and think, "How did I pay this much and barely touch the principal?" That feeling is rational. It's amortization-the schedule that dictates how much of each payment goes to interest versus principal.

In the early years of a typical fixed-rate mortgage, your loan balance is at its peak. Interest is calculated on that big number, so the interest portion of your payment is big too. The system isn't cheating you; it's just doing exactly what you agreed to: charging interest on what's outstanding.

Here's a simplified example. Suppose you have a $300,000 mortgage at 6.5% and your monthly principal-and-interest payment is about $1,896 (numbers rounded for illustration). The first month's interest is approximately:

$$\$300{,}000 \times \frac{0.065}{12} \approx \$1{,}625$$

That means only about:

$$\$1{,}896 - \$1{,}625 = \$271$$

reduces the principal in month one.

That's not a moral failing. It's a math reality: when the balance is huge, interest is huge. As the balance slowly shrinks, the interest portion shrinks, and more of your payment starts hit-ting principal. It's like pushing a boulder that gets lighter one pebble at a time.

Two implications matter for your wealth:

- The *timing* of extra payments is powerful. An extra $200 toward principal in year one is not the same as $200 in year twenty, because it prevents interest from being charged on that $200 for years.
- The "cheap" looking monthly payment isn't the cost of the house; it's the cost of the house *plus a long rental agreement on money.*

If you want proof that small timing changes matter, compare two borrowers with identical mortgages. One makes a single extra $200 principal payment every month for the first five years. The other waits and starts the extra payments in year six. The first borrower doesn't just pay off earlier-they reduce the interest meter earlier, which changes the entire trajectory.

This is why people who learn the math get almost superstitious about principal reduction early on. They're not being dramatic. They're being literate.

The Minimum Payment Trap: the nicest sentence on a credit card statement

Credit card minimum payments are engineered to keep you calm. "Minimum payment due: $___" sounds like a small hurdle. Pay it, and you feel responsible. The lender gets paid. You avoid late fees. Life goes on.

But the minimum payment is not a repayment plan. It's a compliance fee that keeps the account in good standing while interest quietly harvests your future cash flow.

Most credit cards calculate the minimum as a percentage of the balance (often around 1%—3%), sometimes with a small floor amount. Here's the core problem: if your minimum payment is close to the monthly interest, you're barely reducing principal. And as long as principal remains high, interest remains high. It's a self-sustaining loop.

A concrete example makes the trap obvious.

Case study: "Maya and the $5,000 card"

Maya had a $5,000 credit card balance from a rough stretch: a car repair, a few grocery runs that turned into a habit, and one "I'll deal with it later" vacation charged to future Maya. The APR was 24%. The minimum payment was set at 2% of the balance, with a $25 floor.

Maya wasn't irresponsible. She was busy, tired, and trying to keep her head above water. So she paid what the statement asked for: the minimum.

At the start, 2% of $5,000 is $100. And at 24% APR, the interest is also about $100 for the month. That means Maya's first minimum payment essentially *paid rent on money* and did almost nothing else.

Now here's the kicker: as the balance drops by a few dollars, the minimum payment drops too, which keeps Maya pinned in slow motion. The lender isn't forcing Maya to stay in debt; it's simply offering the path of least resistance and charging for the privilege.

When Maya finally looked at the numbers, she switched to a fixed $200 payment-same balance, same APR, but a different rule: the payment wouldn't shrink just because the balance shrank. That one change cut years off her payoff and saved a shocking amount of interest. Not because she found a secret hack. Because she stopped walking on the treadmill.

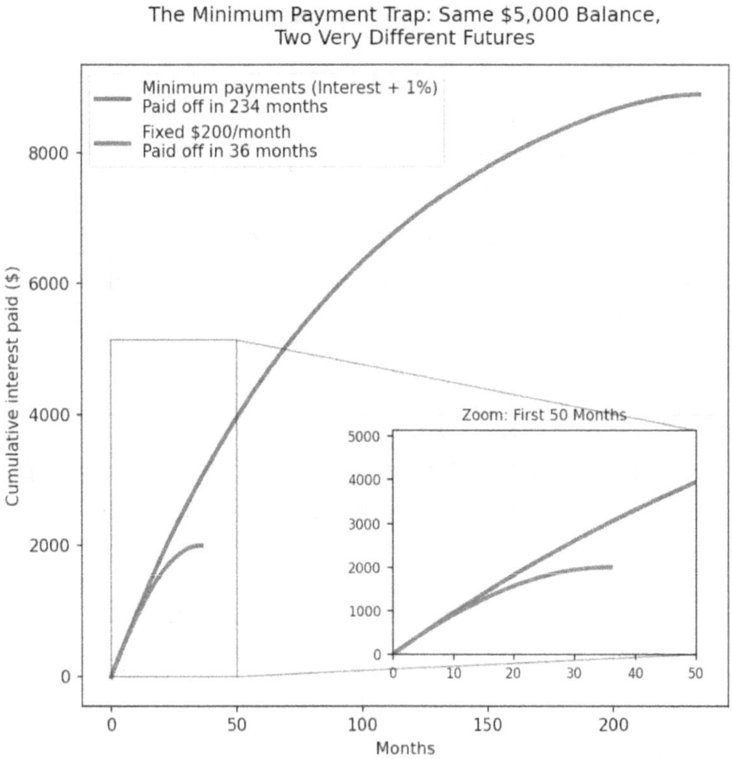

The point of that chart isn't precision down to the penny; it's the shape of the disaster. Minimum payments tend to create a long, grinding tail where you keep paying interest far longer than you'd ever choose if you were forced to watch the clock.

A fixed payment, even a modest one, changes the geometry. It forces principal reduction early, which shrinks the interest meter, which accelerates payoff, which frees cash flow, which builds net worth. That chain reaction is what debt payoff really is: not a moral redemption arc, but a mechanical conversion of monthly cash flow from *interest* to *ownership*.

Why minimum payments feel reasonable (and why that feeling is expensive)

Minimum payments exploit three normal human instincts:

- *We anchor to what's asked.* If the statement says $97, paying $97 feels like "the right amount," even if it's mathematically ridiculous.
- *We avoid pain now.* A bigger payment feels like a loss today, while interest is an invisible loss tomorrow.
- *We underestimate long timelines.* "A few years" sounds like an adult amount of time until it's your life, your cash flow, and your options being narrowed month after month.

The antidote is not willpower. It's changing the default. You don't need to "be better." You need a payment rule that refuses to let the payment shrink just because the balance shrinks.

Borrowing math in the real world: the ``interest cascade" effect

Debt rarely stays neatly contained inside a single account. One high-interest balance can trigger an interest cascade-secondary costs that show up in places you don't mentally connect to the original debt.

Here's how it plays out:

- High utilization on credit cards (using a large percentage of your available credit) can make you look riskier to lenders.
- Looking riskier can raise the APR you're offered on the next loan (or insurance pricing in some contexts), which raises your cost of living.
- A higher cost of living reduces your monthly margin, which makes it harder to pay down the original high-interest balance.

It's not just that interest costs money. Interest reduces flexibility. It reduces choices. It forces you to finance emergencies instead of absorbing them. It makes "unexpected" expenses feel catastrophic because you're already carrying a silent monthly tax.

This is why two households with the same income can have radically different financial lives. One has margin; the other has meters running.

A quick reality check: when ``low payment" is a financial lie

Lenders love to sell monthly payment, because monthly payment is a *vibe*. Total cost is a *verdict*.

A car dealer says, "I can get you into this for $399 a month," as if the universe only bills monthly. But what matters is the total of payments plus interest, and how long your cash flow is committed.

If you want a simple gut-check that works on almost any loan offer, ask two questions:

1. "What is the APR, exactly?"
2. "How many months am I paying, exactly?"

Then translate it into your real life: "How many paychecks is this?" "What would I do with that money if I didn't have to send it to a lender?" Put the opportunity cost on the table. Debt loves darkness.

Practical moves that bend the math back in your favor

You don't have to become a spreadsheet monk to win this game. You need a few moves that change your trajectory without requiring a personality transplant.

1) Replace "minimum" with a rule you can live with. If you're paying minimums on a revolving balance, set a fixed payment that is meaningfully above interest and set it on autopay. Fixed payments create momentum because they don't shrink as the balance shrinks.

A rule that works for many people:

$$\text{Fixed payment} = \max(\$200,\ \text{minimum} + \$50)$$

Adjust the numbers to your reality, but keep the spirit: the payment should not be allowed to "relax" just because the balance does.

2) Attack the interest rate before you attack your pride. If your APR is high, the fastest win is often lowering the rate: promotional balance transfers (watch fees and deadlines), calling the issuer for an APR reduction, or consolidating to a lower-rate product if you can do it without extending the problem. The goal isn't elegance; it's reducing the rental fee.

3) Use "principal-first" thinking on installment debt. On mortgages and other installment loans, confirm how extra payments are applied. Many servicers let you pay extra but apply it as "prepayment" (pushing your due date forward) unless you specify "principal reduction." One shrinks the interest meter; the other mostly changes timing.

4) Track one number: interest paid this month. Not the balance. Not the credit limit. The interest dollars. It's hard to stay casual about debt when you see "$187.43" as the monthly rent you paid to borrow money you already spent.

Actionable takeaway: a 20-minute ``Debt Math Reset"

Do this once, today, and you'll stop guessing.

1. Pick one debt-the one with the highest APR (especially any credit card).
2. Write down: balance, APR, minimum payment, and your current payment.
3. Estimate monthly interest: balance $\times \frac{\text{APR}}{12}$.
4. Set a new autopay amount that is at least monthly interest+ \$100 (or any amount that is safely above interest and sustainable).

5. Schedule one calendar reminder for 30 days from now: "Raise payment by $25 if I can."

That's it. No perfect plan required. Just a new rule that forces your money to buy back your future instead of renting it out month after month.

3.2 Decoding Your Credit Score

A credit score feels personal in the way a bathroom scale feels personal: it pretends to be a judgment about *you*, when it's really a snapshot produced by a machine. The machine doesn't know you're careful. It doesn't know you got laid off once. It doesn't know you've never missed rent in your life. It only knows what's on your credit reports and what the model has learned tends to predict risk.

That's good news. It means you can stop negotiating with your feelings and start pulling levers.

The fastest way to get power over your score is to treat it like a game with known rules: you don't argue with the referee-you learn what draws a foul, what earns points, and what gets reviewed on replay.

First, the myth-busting that actually matters

A few credit-score myths live forever because they *feel* true.

Myth: "Checking my score lowers it."
Checking your own score is typically a "soft" inquiry. Soft inquiries don't hurt your score. What can hurt is applying for new credit and triggering a "hard" inquiry—because the

model treats rapid applications as a possible sign of distress. But monitoring yourself isn't a problem; it's a habit of people who win.

Myth: "Carrying a balance helps my score."
This one is expensive folklore. You don't need to pay interest to build credit. What matters is that you use credit responsibly and pay as agreed. Interest is not a membership fee for a good score. It's a bill you can often avoid.

Myth: "I have one credit score."
You have multiple. Different bureaus can have slightly different data, and different scoring models weight things differently. Your "score" is not a single, sacred number—it's a family of numbers that rise and fall based on the same underlying behaviors.

Once those myths are out of your head, you can focus on what moves the needle.

The score is a prediction engine, not a report card

Credit scoring models were built to answer one question: *How likely is this person to pay back borrowed money on time?* Not "Are they financially wise?" Not "Do they have a high income?" Not "Are they a good person?"

That framing changes how you should think. If the model is trying to predict whether you'll pay, then the model is obsessed with:

- evidence you pay bills on time,
- evidence you aren't overextended,

- evidence you can handle credit over time without chaos.

A score is the smoke alarm, not the fire. It reacts to signals: late payments, high credit card balances relative to limits, lots of new accounts, thin credit history. You don't fix a smoke alarm by yelling at it. You fix the conditions that make it scream.

Your credit report is the raw ingredients

Before you try to "improve your score," you want to understand where the score gets its food: your credit reports.

A credit report is a file maintained by a credit bureau that lists things like:

- which accounts you have (credit cards, student loans, auto loans, mortgage),
- when each account was opened,
- the credit limit or original loan amount,
- the payment history (on-time, 30 days late, 60 days late, etc.),
- balances reported,
- collections, judgments (where applicable), bankruptcies,
- inquiries (applications for credit).

The score is a model that reads that file and tries to predict the future. So improving your score is not mystical. It's a combination of (1) fixing the file when it's wrong, and (2) feeding the model better signals.

The five levers that matter (and why two of them dominate your life)

Most mainstream explanations of credit scoring converge on five big categories for FICO-style scoring:

Factor	*What the model is really asking*
Payment history	Do you pay as agreed, consistently?
Amounts owed / utilization	Are you currently leaning too hard on credit?
Length of credit history	Have you shown stability over time?
New credit	Are you suddenly seeking a lot of credit?
Credit mix	Have you managed different types of credit responsibly?

If you remember nothing else: payment history and utilization are the main event. They're the difference between "boring and safe" and "concerning and expensive." Length of history, new credit, and mix matter too, but they tend to be secondary compared to whether you pay on time and whether you look maxed out.

"Gamifying" your credit score doesn't mean doing anything sneaky. It means understanding which behaviors create the strongest positive signals, then setting up your life so those behaviors happen automatically.

Lever #1: Payment history --- the easiest rule and the hardest to undo

Payment history is the blunt instrument. Pay on time and you're rewarded. Pay late and you're punished. The model doesn't care *why* you were late. It cares that you were.

A late payment is a signal of distress, and distress is what scoring models are built to sniff out. Even one missed payment can drag a strong score down. That drop feels cruel because it *is* cruel: the model is allowed to be pessimistic.

The best strategy here is boring, because boring is what the model calls "low risk":

- Put every loan and credit card on autopay for at least the minimum payment.
- If you can't autopay everything, autopay the accounts with the harshest consequences (credit cards first, then installment loans).
- Align due dates with your pay cycle so autopay doesn't collide with a low-balance checking account.

If you've ever missed a payment because you "forgot," treat that as a systems failure, not a character flaw. A calendar reminder is fragile. Autopay is a machine.

What about paying early? Paying early doesn't usually earn extra points, but it can prevent late payments and can help with utilization (the next lever). Think of it as defensive driving: it doesn't win races, it avoids crashes.

Lever #2: Utilization --- the lever with the fastest results

Utilization is the percentage of your available revolving credit that you're using. Revolving credit usually means credit cards and lines of credit. If you have a card with a $10,000 limit and you're showing a $3,000 balance on the report, that's 30% utilization on that card.

Utilization matters because it's a real-time signal. A late payment might be a one-time disaster. High utilization can look like ongoing strain—like you're living on credit.

Here's the part that turns credit scoring into a game with timing:

Credit cards are typically reported at statement time.
Your score doesn't see your spending habits day by day. It often sees the balance that gets reported when your statement closes (or whatever date the issuer reports). That means you can have the same monthly spending, the same monthly payoff behavior, and two completely different credit scores depending on what your balance looks like on the reporting date.

This is why people get confused. They say, "I pay my card in full every month, so why did my score drop?" Because you paid after the statement closed, not before.

The "statement-close" strategy
If you want utilization to look good, you aim to have a lower balance reported. Two simple approaches:

- Make a mid-cycle payment (or multiple payments) so the statement balance is lower.

- Pay the card down a few days before the statement closing date, not just before the due date.

Due date and statement close date are not the same thing. One is about avoiding late fees. The other is about what gets broadcast to the credit bureaus.

How low should utilization be? There's no single magic number, but generally lower is better, and very high utilization can hurt. Think of utilization like how out of breath you look while carrying groceries. You can be strong and still look strained if you're wobbling. The model can't see your muscles; it sees the wobble.

A real-world case study: the 38-point surprise

Jordan was trying to refinance an auto loan. His income was stable, he had never missed a payment, and he paid his credit card in full every month. The refinance quote came back worse than expected because his score was lower than he thought it should be.

When we looked at the pattern, the reason was almost funny: Jordan was putting nearly all his monthly spending on one rewards card. Around the time his statement closed, his balance would spike to about $4,500 on a $6,000 limit—75% utilization. He would then pay it off a week later, before the due date. Financially responsible, score-wise chaotic.

He made one change: he paid $3,500 of the balance *before* the statement closed, leaving about $1,000 to report (roughly 17% utilization). He didn't spend less money. He didn't carry a balance. He didn't pay interest.

Over the next cycle his score jumped meaningfully, and the refinance math improved. Same person, same income, same debt payoff behavior—just better timing of what the model got to see.

That's what "gamifying" looks like: you learn the scoreboard.

Lever #3: Length of history --- the patience tax

Length of credit history is the category that makes high-achievers feel powerless. You can't grind it. You can't hustle it. You mostly earn it by not doing anything dumb for a long time.

The model likes stability. Older accounts are evidence that you've had access to credit and didn't blow yourself up. This is why closing an old credit card can sometimes hurt: it may reduce your average age of accounts and can also shrink your total available credit (which can raise utilization).

That doesn't mean you should keep every card forever. It means you should be intentional:

- If a card has no annual fee and is one of your oldest accounts, keeping it open can help.
- If a card has an annual fee and you're not getting value, consider downgrading to a no-fee version rather than closing outright (if the issuer allows it).
- If you're tempted to close accounts for "tidiness," remember that neatness isn't a scoring category.

You're not building a minimalist wallet. You're building a track record.

Lever #4: New credit --- the scent of urgency

New credit includes hard inquiries and recently opened accounts. The model gets suspicious when you appear to be rapidly taking on new debt. Sometimes that suspicion is unfair—maybe you're rate-shopping for a mortgage or replacing a broken car. But the model is designed to predict patterns, and rapid seeking can correlate with financial stress.

Practical rules that keep you out of trouble:

- Don't apply for multiple new credit cards in a short span right before you plan to get a mortgage, auto loan, or major financing.
- If you're shopping for a major loan, do your applications within a tight window (often scoring models treat certain rate-shopping inquiries as one event, depending on the model and type of loan).
- When you open a new credit card, expect a small, temporary dip from the inquiry and the lower average account age—then focus on keeping utilization low and payments perfect.

"New credit" is not evil. It's just a signal. The trick is choosing when you can afford the signal and when you need to look boring.

Lever #5: Credit mix --- a supporting actor, not the star

Credit mix refers to having experience with different types of credit—revolving accounts (credit cards) and installment loans (student loans, auto loans, mortgages). The model can treat a person who has successfully managed both types as slightly less risky than someone with only one.

But mix is not worth chasing for its own sake. Taking on a loan you don't need to "build credit" is like buying a treadmill to "build health" and then using it as a coat rack. Don't pay interest for points.

If you have a credit card and one installment loan you're paying on time, you're typically fine. The dominant levers remain: pay on time, keep utilization healthy, avoid frantic credit-seeking.

The most underrated move: make your credit report correct

People talk about raising scores as if it's only about behavior. Behavior matters, but data quality matters too. If your report is wrong, the model will confidently punish you for a life you didn't live.

A practical habit: pull your credit reports periodically and scan for:

- accounts you don't recognize (possible identity theft),
- late payments you believe are incorrect,
- incorrect credit limits (can distort utilization),
- duplicate collections or collections that should have aged off,
- addresses or employers that aren't yours (sometimes harmless, sometimes a clue).

Disputing errors can feel intimidating because it's paperwork and waiting and "customer service." But the payoff can be immediate and tangible: a corrected negative item can raise your score faster than months of perfect behavior.

Treat it like cleaning a window. You're not changing the landscape—you're removing the grime that makes the landscape look worse.

A score-friendly card routine that doesn't require obsession

You don't need to watch your score like a stock ticker. A simple routine captures most of the benefit:

1. Autopay at least the minimum on every account. No exceptions.
2. Use credit cards for normal spending only if you can pay them off reliably.
3. Mid-month (or one week before statement close), make a payment that keeps reported utilization modest.
4. Pay the statement balance by the due date to avoid interest.

It sounds almost too ordinary. That's the point. The scoring system is optimized for ordinary stability. Your goal is to look like the least exciting borrower in the world.

Actionable takeaway: a 30-minute ``Score Control" checklist

Do this once and you'll know exactly what to fix.

1. Find the statement closing date for your primary credit card (not just the due date). Put it on your calendar.
2. Two business days before that close date, pay your balance down so the reported utilization is comfortably low relative to the credit limit.
3. Turn on autopay for at least the minimum payment on every credit account.
4. Pull your credit reports and list any item you don't recognize or believe is inaccurate. Start disputes for the clear errors first.
5. If you plan to apply for a major loan in the next 6–12 months, freeze new credit card applications unless there's a compelling reason.

The score will do what it always does: follow the signals. When you control the signals, you control the story your credit report tells about you—quietly, consistently, and without paying a dollar of unnecessary interest.

3.3 Triage: Good Debt vs. Bad Debt

Debt advice often fails for one simple reason: it treats every dollar you owe as equally urgent. That's like calling every cough a medical emergency. Sometimes you need a cough drop. Sometimes you need the ER. If you throw your money at balances at random—whatever bill is loudest, whatever lender calls the most, whatever makes you feel the most guilty—you can stay "busy" for years while your net worth keeps bleeding out.

Triage fixes that. Triage is the discipline of sorting first, acting second. It turns debt payoff from a panic reaction into a controlled burn: you decide which fires must be put out immediately, which can be contained, and which are safe enough to leave burning while you deal with the real threat.

The goal isn't to win a moral contest called "debt-free." The goal is to stop high-cost debt from eating your future cash flow so you can build wealth with what's left.

Good debt vs. bad debt is too simple---use ``toxic, manageable, strategic'' instead

"Good debt" and "bad debt" are catchy labels, but they hide the real variables. A mortgage at a low rate might be perfectly fine for one person and a financial anchor for another. A student loan might be a bridge to higher income—or a weight that follows you for a decade because the degree didn't pay off. Even credit card debt isn't always "bad" in the moral sense; sometimes it's the least-bad option during a crisis. Labels don't help you decide what to do *today*.

A more useful framework is three buckets:

- *Toxic debt:* debt that compounds aggressively, punishes mistakes, and tends to spiral (usually high-interest consumer debt).
- *Manageable debt:* debt with a defined payoff schedule and tolerable cost, where the main risk is cash-flow strain, not compounding doom.

- *Strategic debt:* debt that supports an asset or outcome with a plausible upside, strong consumer protections, and a cost that doesn't choke your future.

These aren't permanent identities. A debt can move buckets when rates change, income changes, or the loan's terms change. Triage isn't a one-time judgment. It's a living sorting system.

The three-question test that classifies almost any debt

You can triage nearly any debt by answering three questions. Not philosophical questions—numbers-and-contract questions.

1) What is the true cost? (APR and fees)

APR is the "rental fee" for money. It's the best single snapshot for comparing the cost of different debts, especially when products have fees that effectively raise the price. A 29% APR credit card is a very different animal than a 5% student loan, even if the card balance is smaller. Your emotions want to attack the big number. Your net worth wants you to attack the expensive number.

Fees matter because they can turn "reasonable" debt into a trap:

- A balance transfer can help, but a transfer fee changes the math.
- Payday loans can look small on paper until you translate the fee structure into annualized cost.
- "No interest if paid in full" financing can be fine until one missed deadline triggers retroactive interest.

2) How dangerous is the structure? (revolving vs. installment, penalties, and optionality)

Two debts with the same APR can be different levels of dangerous because of structure.

Revolving debt (like credit cards) is open-ended. There may be no fixed payoff date. Minimum payments can keep you on the treadmill for years. Rates can be variable. Your payment can shrink while interest keeps charging. Revolving debt is a flexible tool—until it becomes a permanent tax.

Installment debt (like mortgages, auto loans, many student loans) has a schedule. If you make the payment, the loan amortizes. The structure itself pushes you toward payoff.

Then there are penalty mechanics:

- How harsh are late fees?
- Does the interest rate jump after a missed payment?
- Is the debt secured by something you can lose quickly (car repossession timelines can be unforgiving)?

Optionality is the hidden factor. Optionality means: if life punches you in the face, can you adjust without the debt detonating? Credit cards have flexibility but brutal compounding. Mortgages can be steady but hard to unwind. Student loans can have protections but can also follow you for a long time. Triage is about which debt is most likely to become catastrophic *when life happens*.

3) What did the debt buy? (a consumable, a tool, or an asset)

Debt used for consumption is the easiest to hate because it's literally borrowing from tomorrow to pay for yesterday. But triage is not about shame; it's about outcomes.

A good rule:

- If the debt bought something that is *gone*, it must be cheap and short-lived—or it's toxic.
- If the debt bought something that helps you earn, live safely, or build equity, it might be manageable or strategic.
- If the debt bought something that can be sold near its value (a liquid asset), it can sometimes be defused.

A mortgage (asset) isn't automatically strategic. If the payment consumes your margin and forces you into credit card debt, the mortgage is functionally toxic *in your life*. Meanwhile, a small 0% promotional balance on a credit card can be manageable if you have a plan and the cash flow to kill it before the promo ends. The bucket isn't about the lender's name; it's about the interaction between the loan and your cash flow.

The Debt Triage Map: where common debts usually land

Most people don't need a novel; they need a map they can apply in five minutes. Here's a practical classification, with the understanding that rates and terms can move things around.

Usually toxic

- Credit cards carrying a balance at high APR
- Payday loans and cash advances

- Rent-to-own contracts with punishing effective APRs
- "Buy now, pay later" plans that trigger fees or interest when missed
- Any debt where one mistake causes a giant cost jump

The defining feature isn't that these are "bad." It's that they're optimized to extract money from time. They're compounding machines.

Usually manageable

- Auto loans at a reasonable rate, with a stable payment you can afford
- Student loans at moderate rates, especially with workable repayment options
- Personal loans with fixed terms (assuming the APR is not credit-card level)

Manageable debt is debt that doesn't try to multiply behind your back. It can still be harmful if it squeezes cash flow too tightly, but it generally won't metastasize purely from its structure.

Sometimes strategic

- A mortgage that fits your budget and builds equity over time
- A business loan with clear, measurable ROI and risk controls
- Student loans that plausibly increase lifetime earnings relative to the cost

Strategic debt has a story that makes sense in numbers. It's not "I hope this works." It's "If the plan works, the upside dwarfs the cost, and if the plan doesn't work, the downside is containable."

Tax benefits: the mirage that makes people tolerate terrible deals

Tax advantages are real in some situations, but they're often oversold in everyday conversation. Someone will say, "Don't pay off the mortgage—it's tax-deductible," as if the IRS is reimbursing your interest.

A deduction is not a discount. A deduction reduces taxable income. If you're in a 22% marginal federal bracket, a $1,000 deduction might save around $220 in federal tax. You still paid $1,000 in interest to "save" $220. That may still be fine if you value other benefits (stability, homeownership, fixed payment), but the tax angle should never hypnotize you into ignoring the price.

The same goes for student loan interest deductions and other conditional benefits: they can help at the margin, but they don't transform expensive debt into cheap debt. The APR is still the rental fee, and the rental fee still drains cash flow.

Use tax rules as a *tiebreaker*, not a foundation.

A concrete case study: when paying ``the smallest bill'' keeps you broke

Elena had four debts:

Debt	*Balance*	*APR*	*Min. Pmt*	*Notes*
Credit Card A	$2,300	29%	$70	Rewards card, carried balance after a move
Credit Card B	$7,800	21%	$210	Older card, higher limit, always "manageable" until it wasn't
Auto Loan	$11,500	6.4%	$320	Needed for commuting, payment fits budget
Student Loan	$18,900	4.9%	$205	Fixed payment, no drama

Elena's instinct was to pay off Credit Card A first because it was the smallest. That instinct wasn't irrational—it promised quick emotional relief. But her real emergency wasn't the smallest balance. Her emergency was the interest meter.

Here's what triage revealed:

- Credit Card A and B were toxic because of high APR and revolving structure.
- The auto loan was manageable: fixed term, tolerable rate, asset she needed to earn.
- The student loan was manageable-to-strategic depending on her career income (and it had a lower rate with predictable terms).

So the plan wasn't "pay everything a little extra." That's how people stay stuck. The plan was: keep every account current (minimums everywhere), then pour every extra dollar into the *most toxic* debt first, because that's where each extra dollar buys the most future.

What happened in real life is the point: within a couple of months of directing her extra payments to Card A (29%), her total interest charges fell noticeably. Not because she became more disciplined, but because she removed the most expensive compounding first. When Card A died, the freed payment rolled into Card B. Her score started to improve as utilization dropped, which opened the door to refinancing options she didn't have when she looked maxed out.

Elena didn't need more motivation. She needed a target worth shooting.

The ``panic-payment" pattern (and how to replace it)

Panic-paying looks like this:

- You throw $50 at a medical bill because it feels urgent.
- You throw $75 at a credit card because the app is sending notifications.
- You throw $100 at a store card because it's embarrassing.
- You run out of cash, then you use the credit card again.

The pattern creates motion without progress. It also quietly protects the most profitable debt for the lender: high-interest revolving balances that stay alive.

A triage-based system replaces panic with two rules:

1. *Stabilize:* pay minimums on everything to avoid late fees, credit damage, and penalty APRs.
2. *Eliminate:* direct every extra dollar to the single most toxic balance until it's gone.

This isn't even about being "optimal" in some abstract sense. It's about stopping the debt that can spiral from a mild problem into a crisis.

Refinancing and consolidation: triage tools, not magic

Refinancing can move a debt from toxic to manageable by lowering the APR or converting revolving debt into a fixed payoff schedule. Consolidation can simplify your life. But neither is automatically good. They are tools. Tools can build a house or break a window.

A consolidation loan that lowers a 24% credit card to 12% can be a real win. But two failure modes are common:

- You consolidate the card debt, then keep using the cards, and now you have *both* the loan and the new card balances.
- You stretch the term so long that the lower payment feels good but the total interest paid balloons.

Triage tells you when refinancing is a priority: when the current debt is toxic *and* you have a realistic path to stop adding to it. If spending habits are still actively inflating the balance, lower APR helps, but it won't save you by itself.

A blunt but useful ranking system: the Debt Danger Score

If you want something you can apply quickly without turning your kitchen table into an accounting firm, give each debt a simple "danger score" from 1 to 10 based on four signals:

- *APR level:* high rate = higher danger.
- *Structure:* revolving = higher danger than fixed installment.
- *Penalty risk:* harsh late fees/penalty APR/repo risk = higher danger.
- *Cash-flow strain:* if the payment squeezes your essentials, danger rises.

Then sort by danger, not by balance size. If two debts are close, prioritize the one that is most likely to spiral. This is triage in plain language: you treat the wound that can kill you first.

Actionable takeaway: build your personal triage list in 25 minutes

Set a timer. Do it messy. Done beats perfect.

1. Write down every debt with: balance, APR, minimum payment, and whether it's revolving or installment.
2. Put a star next to any debt with APR above 15% and any revolving balance you're carrying.
3. Circle any debt where one mistake triggers severe penalties (penalty APR, repossession risk, deferred-interest deadline).

4. Label each debt: *toxic, manageable,* or *strategic,* based on cost and structure, not vibes.
5. Choose one toxic debt as your "fire." Set autopay minimums on everything else. Every extra dollar goes to the fire until it's out.

If you do only that, you'll stop paying the "confusion tax"—the money lost because you were working hard with no system. Triage turns your cash flow into a weapon instead of a bandage.

3.4 Strategies for Debt Elimination

Debt payoff advice is usually delivered like a commandment: *pay it off.* "That's like telling someone to "get fit." True, but useless." The difference between a plan that works and a plan that becomes another abandoned tab in your browser is not math—it's *momentum.* You need a method that creates wins fast enough to keep you going, while quietly doing the heavy lifting in the background.

Two strategies dominate for a reason. They're simple, they're battle-tested, and they force your money to behave with intention:

- *Debt Avalanche:* pay extra toward the highest interest rate first (math-first).
- *Debt Snowball:* pay extra toward the smallest balance first (motivation-first).

Both approaches share a foundation that matters more than the brand name: you keep all debts current (minimum payments on everything), and you concentrate every extra dollar on one target at a time until it's eliminated. That focus is the secret sauce. Without focus, your extra payments get diluted into "a little bit everywhere," which feels responsible but often keeps you stuck.

3.5 Step zero: a payoff plan starts with *stability*, not heroics

Before the strategy, there's stability. Debt payoff collapses when one surprise expense sends you back to the card you're trying to kill. So the first move is not "maximize payments." The first move is "stop new damage."

Stability usually means:

- minimum payments automated so you don't trigger late fees or penalty rates,
- a tiny cash buffer (even $500—$1,000) so your plan doesn't shatter the first time your tire goes bald,
- a realistic monthly "attack amount" you can sustain without starving your life.

If you're already in the danger zone—collections calls, multiple accounts past due—your "strategy" is different. The immediate priority becomes getting current, negotiating where needed, and preventing the kind of credit damage that creates years of expensive borrowing. But for the common situation (a few cards, maybe a car loan, maybe student debt), Snowball and Avalanche are the workhorses.

3.6 Debt Avalanche: the math-first strategy that saves the most money

Debt Avalanche is ruthlessly logical. You list debts by interest rate from highest to lowest. You make minimum payments on everything. Then you throw every extra dollar at the highest APR debt first. When it's gone, you roll that freed-up payment into the next-highest APR, and so on.

Why it works is almost insultingly simple: a dollar of principal paid on a 29% card prevents far more future interest than a dollar paid on a 6% auto loan. In other words, Avalanche buys the most interest relief per dollar.

If your brain craves efficiency, Avalanche feels like taking control of a chaotic room and finally putting everything in the correct drawer. It's clean. It's elegant. And on paper, it often wins.

But paper doesn't make the payments—you do.

3.7 Debt Snowball: the motivation-first strategy that keeps you in the game

Debt Snowball ignores APR and targets the smallest balance first. The mechanics are the same: minimums everywhere, extra toward the target, then roll the freed payment to the next smallest balance.

Snowball is built on a psychological truth: early wins change how you see yourself. A paid-off account is a visible trophy. It reduces the number of bills, the number of lenders, the number of mental tabs open in your head. That reduction has power.

Snowball is also surprisingly practical for messy real life. Small balances often come with annoying store cards, random medical bills, or lingering accounts that you keep "meaning to deal with." Knocking them out clears clutter and can reduce the chance you miss a due date.

Here's the trade-off: Snowball can cost more in interest if it delays paying down a very high-rate debt. It's not mathematically optimal. It's behaviorally optimized.

And behavior is not a side issue. For many people, behavior *is* the whole problem.

3.8 The uncomfortable truth: most people don't pay debts the way math says they should

If everyone naturally followed Avalanche, we wouldn't need debt payoff books. In reality, many borrowers spread extra payments across debts in a way that feels "fair," like dividing a pizza-each balance gets a slice. Researchers have observed a "balance-matching" tendency: people allocate extra payments roughly in proportion to each balance rather than concentrating payments where they save the most interest.

It's understandable. Concentration feels risky. It feels like neglecting other bills. But in debt payoff, concentration is what creates escape velocity. Paying an extra $25 on five different accounts often produces five tiny, invisible non-wins. Paying an extra $125 on one account produces a win you can *see*.

Snowball and Avalanche both correct the same human error: dilution.

3.9 A concrete example: same debts, same budget, different outcomes

Suppose you have three debts:

Debt	*Balance*	*APR*	*Min. Pmt*	*Notes*
Card A	$600	18%	$25	Small balance, annoying to track
Card B	$2,400	27%	$80	High APR, main interest engine
Loan C	$8,000	7%	$160	Installment loan, fixed term

Assume you have $250/month to put toward debt elimination *on top of minimums*. That $250 is your "attack amount."

Avalanche order: Card B (27%) → Card A (18%) → Loan C (7%).
Snowball order: Card A ($600) → Card B ($2,400) → Loan C ($8,000).

Mathematically, Avalanche saves more interest because it attacks the 27% fire first. But psychologically, Snowball gives you a win fastest: Card A disappears quickly, and suddenly one bill is gone. Your brain gets a hit of progress.

This is where people choose incorrectly—by trying to choose a universally "best" method. The best method is the one you will execute for months without quitting.

3.10 Which one should you choose? Use a personality-based decision rule

Instead of asking, "Which strategy is best?" ask, "What causes me to quit?"

Choose *Avalanche* if:

- you stay motivated by efficiency and numbers,
- you can tolerate a long first phase without a visible win,
- high-interest debt is a significant part of your situation.

Choose *Snowball* if:

- you need quick wins to believe change is real,
- you feel overwhelmed by the number of accounts,
- you've started payoff plans before and abandoned them.

There's also a hybrid approach that works beautifully for real humans: *Snowball a tiny "starter win," then Avalanche the rest.* Pay off one small balance to get momentum, then switch to attacking the highest APR. This can be the best of both worlds: you get psychological traction without sacrificing too much interest savings.

3.11 The rollover principle: how debt payoff accelerates

Both methods rely on the same engine: rollover. When one debt is paid off, you don't "reward yourself" by lowering your total monthly payment. You roll that freed payment into the next target.

This is where payoff starts slow and then suddenly feels like it's happening *to you* rather than *because of you*. Your attack amount grows without you earning a raise. Each payoff increases your monthly firepower.

Rollover is compounding—but in your favor.

3.12 The visual: Snowball vs. Avalanche order (and the rollover effect)

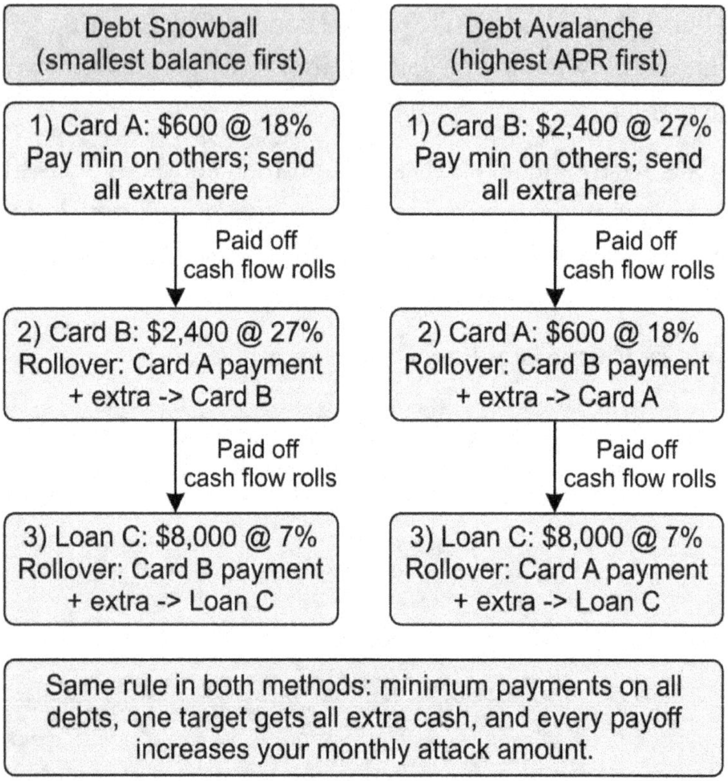

3.13 A real-world case study: when the ``best" plan fails, and the ``sticky" plan wins

Marcus had good income and a messy debt stack: two credit cards, a store card, and a personal loan. He was the kind of person who could understand Avalanche immediately and even built a spreadsheet for it. Then he didn't follow it.

What happened wasn't a mystery. His highest APR card also had the biggest balance. Month after month, he threw extra money at it, watched the balance barely move, and felt like he was shoveling snow in a blizzard. Meanwhile the store card kept nagging him with emails and the small balance felt like a splinter in his brain.

So he switched to Snowball for one reason: he needed proof that the plan worked. He killed the store card first, then the smaller credit card, then attacked the big high-APR balance with a much larger rollover payment. Was it mathematically perfect? No. Did he actually finish? Yes.

That's not a cute story. It's the reality of debt payoff: the winning strategy is the one you can stand.

3.14 When you need help: Debt Management Plans and the scam minefield

Sometimes the problem isn't strategy—it's that the interest rates are so high and the budget is so tight that even a perfect strategy feels like bailing out the ocean with a mug. If you're current on payments but drowning, a legitimate nonprofit credit counseling agency may be able to set up a Debt Management Plan (DMP). In a DMP, you typically make one monthly payment to the agency, and they pay creditors. Creditors may reduce rates or fees in exchange for predictable repayment.

A DMP isn't magic, and it's not for everyone. It can involve closing credit card accounts, and it requires steady payments. But for the right person, it turns chaos into a lane.

On the darker side, the debt-relief space is full of traps: companies that charge large upfront fees, promise to "erase" accurate negative information, or pitch settlement strategies without explaining the consequences (collections calls, credit score damage, potential tax impacts, lawsuits). If the pitch sounds like a cheat code—*"we can remove your debt without pain"*—assume you're the product.

The safe posture is simple: if you're paying someone for help, be able to explain in plain language what they do, what it costs, what the risks are, and what happens if you stop.

3.15 A practical payoff blueprint you can run on autopilot

Regardless of Snowball or Avalanche, the operational system is nearly identical:

1. List all debts with balance, APR, minimum payment, and due date.
2. Set autopay minimums for everything (protect your payment history).
3. Choose your target debt using Snowball or Avalanche.
4. Set a fixed monthly "attack amount" that goes to the target.

5. Each time a debt is paid off, roll its payment into the next target *the very next month*.

The key is to turn the plan into a default behavior. If you have to re-decide every month, you'll eventually negotiate with yourself. Autopilot doesn't negotiate.

3.16 Actionable takeaway: pick a method today, then run a one-month test

Make this decision small enough that you'll actually do it.

- If you care most about saving money and you can tolerate delayed gratification, run *Avalanche* for 30 days.
- If you care most about momentum and you've quit before, run *Snowball* for 30 days.

At the end of the month, don't ask, "Which method is superior?" Ask, "Which method did I follow without drama?" Then commit for six months and let rollover do what it does best: turn a tight budget into increasing leverage.

The day you pay off your first debt and redirect that payment elsewhere, your financial life changes in a specific, measurable way: your monthly obligations shrink, your margin grows, and the future starts costing less to access. That is the feeling you're chasing—not perfection, but relief you can calculate.

Chapter 4

Investing Made Simple

Think of your money as an employee. If it's sitting in a checking account, it's sleeping on the job. Investing is simply the act of putting that money to work so it earns a paycheck of its own. You don't need a PhD or six screens to succeed; you just need to understand a few timeless mechanics that turn small habits into massive freedom.

4.1 How Financial Markets Actually Work

If the stock market feels like a casino, it's usually because you're picturing the loudest part: the flashing quotes, the screaming headlines, the chart that looks like a heart monitor. Casinos are designed to *separate you from your money* with entertainment and short-term adrenaline. Financial markets, at their core, are designed to *allocate money*—quietly, relentlessly—toward businesses and projects that (we hope) will use it productively. The difference isn't moral; it's mechanical. A roulette wheel can't build a better toothbrush. A company can.

A useful way to see the machinery is to start with a simple question: *Why do stocks exist at all?* Because businesses need cash to grow. Some businesses borrow it. Some sell ownership to get it. Markets are the meeting place where those choices become tradable contracts-ownership claims and IOUs-so millions of strangers can fund real work without ever meeting.

4.2 The Market Isn't One Place-It's Two Different Machines

Most people imagine "the market" as a single arena where money sloshes around. In reality, there are two distinct systems with two distinct jobs.

Machine #1: the primary market. This is where money goes *to* a business. When a company first sells stock to the public (an IPO) or later sells additional shares (a secondary offering), investors' dollars help fund payroll, factories, research, acquisitions-whatever the business claims it will do with the capital.

Machine #2: the secondary market. This is where investors trade those existing shares with each other. When you buy shares of a giant public company on a normal Tuesday, your money usually goes to another investor, not to the company itself.

That second machine can sound pointless until you realize its superpower: it creates *liquidity*, the ability to turn an investment into cash without begging someone to take it off your hands. Liquidity is oxygen. Without it, far fewer people would be willing to invest in the first place, because "good luck getting your money back" is not a compelling sales pitch.

Here's the subtle but important twist: even when a trade doesn't directly fund a company, the price discovered in the secondary market still affects the company's real-world options. A higher stock price can make it easier to raise money later, pay employees with stock, or use shares to buy other companies. A lower stock price can make all of that harder. So the secondary market is not charity, but it's not a pointless betting parlor either. It's a continuous voting-and-pricing system that shapes what companies can do.

4.3 Stocks vs. Bonds: Owning the Lemonade Stand or Lending It Money

Imagine your friend Maya runs a lemonade stand that's getting unexpectedly popular. She wants a better cart, a bigger cooler, and a sign that doesn't look like it was attacked by a marker. She needs $1,000.

She can get that $1,000 in two broad ways:

Option A: Sell a slice of the business (*equity*). Maya offers you 10% of the lemonade stand for $1,000. Now you are an owner. If the stand booms, your 10% becomes more valuable. If profits grow, you might get some share of them (in real companies, that's dividends or reinvested earnings that raise future value). If the stand flops, you don't get a guaranteed refund. Ownership is a variable-reward deal.

Option B: Borrow the money (*debt*). Maya gives you an IOU: she'll pay you 6% interest and return your $1,000 in a year. Now you are a lender. You don't get 10% of the upside if the stand becomes the next beverage empire. You get your agreed interest-if she can pay. Lending is a fixed-reward deal with specific promises attached.

That's the basic difference between *stocks* (equity, ownership) and *bonds* (debt, lending). Every complicated investing conversation eventually collapses into this: do you want a claim on *profits* (and uncertainty), or do you want a claim on *payments* (and the risk they won't happen)?

The casino stigma often comes from treating stocks like lottery tickets-something you buy hoping someone else will pay more later. But the cleanest mental model is older and calmer: a stock is a piece of a business's future cash flows. The price may bounce around, but the underlying idea is almost boring: you own a small claim on whatever the business becomes.

4.4 What a Stock Price Actually Is (And Why It Moves)

A stock price is not a judge's verdict on whether a company is "good." It's not even a precise measurement of value. It's simply the most recent agreement between a buyer and a seller about what one share is worth *right now*, given what they believe, what they fear, and what they need.

Two forces are always wrestling inside that number:

- *The business engine:* revenue, costs, competition, innovation, management execution-things that determine how much cash the company might generate over time.
- *The mood engine:* interest rates, inflation worries, recession fears, hype cycles, forced selling, short-term news, and the basic fact that humans panic in groups.

When people say, "the market is irrational," they're usually noticing the mood engine. When people say, "the market is efficient," they're usually pointing at the business engine over longer stretches. Both observations can be true, depending on your time horizon. Over days and weeks, prices can behave like gossip. Over decades, they tend to behave more like arithmetic.

This is where the casino analogy quietly breaks. In a casino, the house builds the edge into the rules. In markets, there is no guaranteed "house" extracting expected value from every trade in the same way. There are frictions-fees, spreads, taxes, bad timing-but the system is not structurally designed so that *you must lose*. You can make money simply because businesses, in aggregate, can create value over time. That's not a loophole. That's the point.

4.5 A Trade Is a Conversation: The Bid, the Ask, and the Spread

If you've ever looked at a stock quote and thought, "Why does it say two prices?" you've just found one of the most practical pieces of market plumbing.

At any moment, there is:

- the *bid*: the highest price someone is willing to pay to buy right now,
- the *ask* (or offer): the lowest price someone is willing to accept to sell right now.

The gap between them is the *spread*. That spread is a cost you often pay without noticing. If you hit "buy now" at the ask and then immediately sell at the bid, you'd lose roughly the spread (ignoring commissions). On a huge, heavily traded stock, the spread might be a penny. On a thinly traded stock, it can be wide enough to feel like a toll booth.

This is one reason markets can feel like a casino to beginners: you can make a perfectly reasonable decision and still see a small loss instantly. That's not the universe punishing you. That's just the mechanics of liquidity.

4.6 Orders: How You Speak to the Market (and How It Talks Back)

Your broker app is not "the market." It's your microphone. The way you place an order determines what you're prioritizing: *speed* or *price control*.

Market order. This says, "Get me in (or out) now." It prioritizes execution. The trade will typically fill quickly, but the exact price can surprise you, especially in fast-moving markets or when trading outside normal hours. You're basically saying, "I accept whatever the current crowd is offering."

Limit order. This says, "I'll buy, but only up to this price," or "I'll sell, but only down to this price." It prioritizes price control. The trade might not happen at all if the market never reaches your limit.

Neither is morally superior. They solve different problems. If you're buying a diversified index fund as part of a long-term plan, you might care more about getting the trade done than squeezing out a few cents. If you're dealing with something illiquid, volatile, or you're trading during a chaotic news moment, price control may matter a lot.

A quiet truth: many painful "I got robbed" stories are really "I used an order type without understanding the trade-off."

4.7 A Concrete Example: The Morning a Headline Hits

Picture a normal weekday morning. A well-known company reports earnings. The headline is dramatic: *"Company X Misses Expectations."* The stock drops sharply in pre-market trading. Social media declares the business dead by lunchtime.

Two investors see the same headline.

Investor A places a market order to sell immediately. In the first burst of panic, liquidity can be thinner and spreads can widen. The order fills fast, but at a worse price than expected—because "sell now" means "sell to whoever is willing to buy right now," and right now the room is full of people backing away.

Investor B pauses and looks for the actual numbers: revenue, profit margins, forward guidance, and whether the miss was a one-time event or a structural problem. Maybe the company had a temporary expense spike but still grew customers. Investor B decides either (1) to do nothing, because the long-term story hasn't changed, or (2) to sell later with a limit order at a price they're willing to accept.

This isn't about being smarter. It's about refusing to let the loudest moment set the terms of your trade.

Markets are constantly processing information, but they also constantly process *emotion*. The casino feeling often comes from walking into the middle of an emotional stampede and assuming the stampede is truth. Sometimes it is. Often it isn't. Either way, you don't want your order type to turn a temporary frenzy into a permanent decision.

4.8 Why Downturns Happen (and Why They're Not System Errors)

If markets only went up smoothly, they would feel less like gambling. They would also be lying to you.

Prices fall for real reasons: recessions reduce profits, interest rates change the math of future cash flows, wars disrupt supply chains, new competitors rewrite the rules. Prices also fall for psychological reasons: fear becomes contagious, leveraged investors are forced to sell, and uncertainty makes people demand a bigger "discount" before they're willing to buy.

A downturn is not a glitch. It's the market doing two brutal but useful jobs at once:

- *Resetting expectations.* When investors realize they were too optimistic, prices adjust.
- *Repricing risk.* When the world feels shakier, people require higher potential returns to take the same risk, which means prices today must be lower.

The uncomfortable implication is also the empowering one: volatility is often the admission price for growth. If you want the long-term returns that businesses can generate, you have to tolerate the short-term fact that humans are not calm creatures.

This is also where personal finance becomes personal. A downturn doesn't just test your portfolio; it tests your *cash flow*. If your monthly budget is tight and your emergency fund is thin, a falling market can force you to sell at the worst moment because you need rent money. Markets punish the investor who needs liquidity on a deadline. They reward the investor who has time.

4.9 A Real-World Case Study: March 2020 and the Difference Between "Price" and "Value"

In late February and March of 2020, markets around the world dropped rapidly as COVID-19 spread and governments imposed shutdowns. The news wasn't subtle: hospitals, layoffs, empty streets, and a global economy slamming the brakes. Stock charts didn't drift downward; they fell like an elevator cable had snapped.

This was a perfect storm for the casino narrative. People saw red numbers and concluded the market was "just gambling." But look at what was actually happening under the hood.

Investors were trying to price an unprecedented set of questions in real time: How long would shutdowns last? Would consumers return? Would companies survive a sudden revenue drought? Would governments step in? Would supply chains recover? In a situation like that, the market can't be stable because stability would imply confidence—and confidence was not available for purchase.

Now comes the key lesson: a collapsing *price* did not automatically mean the productive capacity of every business vanished overnight. Some businesses were genuinely impaired for years. Some were temporarily disrupted. Some became more valuable. The market, in aggregate, eventually recovered as information improved, policy responses became clearer, and the economy adapted. Investors who sold in panic locked in losses. Investors who had the cash flow and temperament to hold (or keep buying diversified funds) benefited from the recovery.

The point isn't that markets always bounce back quickly. The point is that the market wasn't a roulette wheel; it was a messy, human, information-processing machine trying to attach numbers to a foggy future.

4.10 Where ETFs Fit: A Stock-Like Wrapper Around Many Stocks

A common source of confusion is seeing an ETF (exchange-traded fund) trade like a stock and assuming it must be "more speculative." The truth is almost the opposite: an ETF is often used to reduce risk by buying a whole basket at once.

An ETF is a fund that holds a collection of assets-often hundreds or thousands of stocks or bonds-but its shares trade on an exchange during the day. That means you can buy or sell it like a stock, at market prices that can move slightly around the fund's underlying value.

If individual stocks are single trees, an index ETF is buying a whole forest in one transaction. The trading action can look similar, but the exposure is dramatically different. One is a bet on a company. The other is a bet on an economy, a sector, or a broad slice of global business activity.

4.11 The "Fairness" Question: Are You Just Exit Liquidity for Someone Smarter?

This fear is emotionally reasonable: *If someone is selling to me, do they know something I don't?* Sometimes, yes. Often, no.

People sell for countless non-information reasons:

- They need cash for a down payment.
- They're rebalancing a portfolio.
- They're harvesting a tax loss.
- Their fund has redemptions and must raise cash.
- They're reducing risk because of a job change.

And many professional traders aren't "smarter" in the way you mean. They may be faster, more specialized, or operating on time horizons (seconds, minutes, days) that are irrelevant to a long-term investor. If your goal is to fund a life 10, 20, or 30 years from now, you don't need to win a fistfight over pennies today. You need to own productive assets, keep costs low, and stay solvent and steady long enough for compounding to do its work.

The market can still be unfair in pockets. Scams exist. Pump-and-dumps exist. Some corners of the market are thinly regulated or intentionally confusing. But the broad public markets that most long-term investors use are not built to be a rigged carnival game. They are built to be a high-speed system for setting prices and moving capital. The noise is a side effect of the speed.

4.12 A Practical Way to Use This Knowledge Tomorrow Morning

Here's the simplest action that changes your investing life: decide whether you are a *trader* or an *owner* when you place a trade.

If you're acting like an owner:

- You care more about what you own (a diversified basket of businesses) than the day-to-day quote.
- You use orders thoughtfully-often a simple approach is enough, but you avoid making panicked decisions in fast markets.
- You expect downturns as part of the deal and build your cash flow (emergency fund, manageable monthly commitments) so you're not forced to sell at bad times.

If you're acting like a trader:

- You are competing with people who do it full-time, with better tools, faster data, and fewer emotions involved.
- Your biggest enemies aren't just "being wrong." They're costs, spreads, taxes, and impulsive decisions.

A stock market that feels like a casino is often a stock market you're using like a casino: frequent bets, short time horizons, and decisions made under dopamine and dread. Treat it instead like what it is-a public, liquid marketplace for slices of real businesses and real loans-and it becomes less of a thrill ride and more of a tool: imperfect, noisy, sometimes scary, but fundamentally useful for turning patience into ownership.

4.13 Understanding Asset Classes

You can build a life on a reliable paycheck and a sensible budget, but investing is where your money starts doing *night shift* work. The catch is that money has personalities. Some dollars behave like caffeine: jittery, energetic, occasionally brilliant. Other dollars behave like a seatbelt: not exciting, but you notice them most when something goes wrong. Those personalities have names in finance—*asset classes*—and most portfolios are made from just two main ingredients: the gas pedal (stocks) and the brakes (bonds and cash).

The goal isn't to pick the "best" ingredient. It's to cook something you can actually eat for years. A portfolio that's all gas can get you somewhere fast, but it can also spin out on a wet road. A portfolio that's all brakes might keep you safe, but it can leave you stuck in place while time and inflation quietly tow your purchasing power away.

Asset Classes: The Useful Definition (Not the Fancy One)

An asset class is simply a group of investments that tend to behave similarly because they're driven by similar forces. When you hear someone say, "I'm diversified," they often mean, "I own a lot of things." That might help, but the deeper question is: do you own things that *respond differently* to the same economic weather?

Stocks respond to business growth and investor optimism (and sometimes investor mania). Bonds respond to interest rates and the borrower's ability to pay. Cash responds to central bank policy and inflation. Put them together, and you get a portfolio that doesn't rely on one single story always being true.

It helps to think of asset classes like different tools in a toolbox. A hammer is great—until you need to cut something. Stocks are powerful—until you need stability on a deadline.

The Gas Pedal: Stocks (Equities)

A stock is a slice of ownership in a company. Owning stocks is saying, "I want to participate in what businesses can build." Over time, businesses can hire people, invent products, raise prices, lower costs, expand into new markets, and compound their earnings. When that happens across thousands of companies, the stock market becomes a rough mirror of human productivity.

The reason stocks are called the gas pedal is simple: they have historically offered higher long-term growth potential than safer assets, but they can be emotionally violent in the short run. You don't get paid for owning stocks because it's fun. You get paid because it's sometimes terrifying and frequently inconvenient.

Here's what makes stocks *feel* risky:

- Prices can fall 30% or 50% without asking your permission.
- Bad news arrives on a schedule that does not match your rent due date.
- Your timeline may be long; your nervous system is not.

Here's what makes stocks *actually* powerful:

- They are claims on profits. Profits can grow.
- Many companies reinvest earnings to create more earnings.
- Inflation can raise nominal revenues and earnings over time, which can support higher stock prices.

That last point matters more than it sounds. Inflation is the slow leak in your financial tire: you can ignore it for a while, but you will eventually find yourself riding on the rim.

The Brakes: Bonds and Cash

"Brakes" doesn't mean "bad." It means "designed to keep you in control." Bonds and cash do two jobs that stocks are terrible at: *stability* and *timing*.

Bonds (Fixed Income): Lending With Rules Attached

A bond is a loan. You lend money to a government or a company, and in exchange you're promised interest payments and the return of principal at maturity. In everyday language: you're not buying a dream; you're buying a schedule.

Bonds can still be risky—just in different ways:

- *Interest-rate risk:* if new bonds are issued with higher interest rates, older bonds with lower rates become less attractive, so their prices fall.
- *Credit risk:* the borrower might not pay you back (more relevant for lower-quality corporate bonds).
- *Inflation risk:* if inflation runs hot, fixed interest payments buy less in the future.

Even with those risks, bonds often act like emotional shock absorbers in a portfolio. When stocks are dropping because investors are panicking about growth, high-quality bonds sometimes hold steadier—or even rise—because investors are seeking safety and predictable cash flows. "Sometimes" is doing important work there; correlations change. But the broader point stands: bonds are not there to make you rich overnight. They're there to help you avoid making a catastrophic decision at the worst possible time.

Cash: Boring, Liquid, and Occasionally a Superpower

Cash (and cash equivalents like money market funds or Treasury bills) is the least glamorous asset class, which is exactly why it's useful. Cash is what you use when life happens on Tuesday and your portfolio's "long-term plan" is scheduled for 2046.

Cash has two big advantages:

- *Liquidity:* you can use it immediately without selling anything at a bad time.

- *Stability:* its value doesn't swing wildly day to day.

Cash also has a quiet downside: inflation. If inflation averages 3% and your cash earns 1%, you're losing purchasing power every year, even though the dollar balance looks "safe." Cash is safe in the way a parked car is safe: it won't crash, but it also won't take you anywhere.

So cash is not a growth strategy. It's a *resilience* strategy.

Risk and Return: You Don't Get One Without the Other

People often talk about "risk" as if it's a moral flaw: risky is bad, safe is good. Markets don't work like that. In markets, risk is usually a price tag. If an investment is genuinely safer and everyone knows it, the expected return tends to be lower because people are willing to pay more for safety.

That's why the gas pedal and the brakes exist as categories: they trade off against each other.

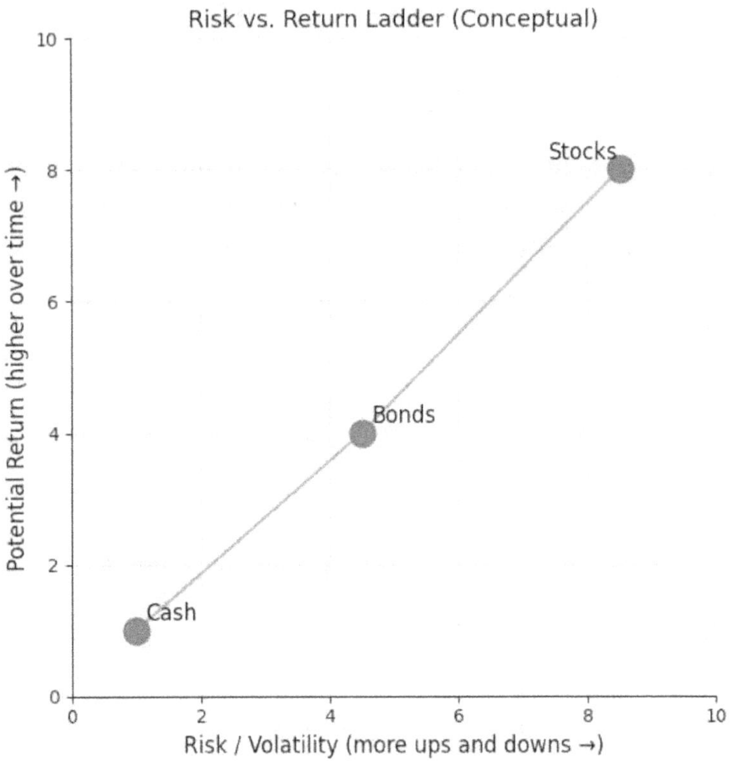

The chart is deliberately simple because the idea is simple: you're not choosing between "smart" and "dumb." You're choosing between different kinds of discomfort. Stocks tend to punish you with volatility. Cash tends to punish you with stagnation. Bonds tend to punish you with disappointment when rates rise or inflation flares. Adult investing is picking the set of punishments you can live with long enough to win.

A Real-World Scenario: The Layoff, the Down Payment, and the Portfolio That Didn't Break

Consider two people, both 35, both investing seriously.

Alex invests every spare dollar into a stock fund. No cash cushion beyond a few weeks of expenses. Alex's logic is clean: "Stocks have the best long-term returns. I'm young. I can handle it."

Jordan invests too, but keeps a real emergency fund and also holds a bond fund inside retirement accounts. Jordan's logic sounds less impressive at parties: "I don't want to be forced to sell when life punches me."

Now imagine a rough year: a recession scare hits, stocks drop, and layoffs roll through Alex's industry. Alex loses their job and needs cash for bills. The portfolio is down, and selling stocks now locks in losses. Alex sells anyway—because the landlord doesn't accept "long-term expected returns" as payment.

Jordan also faces anxiety, but doesn't face a forced sale. The emergency fund covers bills while job searching. The bond allocation helps keep the overall portfolio from dropping as hard. Jordan can keep stock investments intact and, if the job situation stabilizes, may even keep contributing when prices are lower.

Both investors had the same "asset" at the center—stocks—but the outcome was determined by the supporting cast. This is what the brakes are for. They don't exist to maximize your best-case scenario. They exist to keep your worst-case scenario from turning into a permanent financial scar.

The Hidden Third Ingredient: Time Horizon (It Chooses Your Asset Mix)

Asset allocation sounds like math. It's often psychology with a calculator.

A *time horizon* is when you'll need the money. If you'll need it soon, volatility is more dangerous because you have less time to recover. If you won't need it for decades, volatility is less dangerous—*unless* it triggers panic selling.

A practical way to think about it:

- Money needed in the next few years prefers cash and high-quality bonds.
- Money meant for the distant future can lean more heavily into stocks.

This isn't a rule carved into granite; it's a way to prevent a common mistake: investing short-term money in long-term assets. Stocks can be great for retirement. They can be terrible for your house down payment if the purchase date is non-negotiable.

Funds Are Wrappers, Not Ingredients

Many investors get tangled in product names. Mutual funds. ETFs. Index funds. Target-date funds. These are mostly *wrappers*—containers that hold ingredients.

If you buy a total stock market index fund, the ingredient is *stocks*, even though the product is a fund. If you buy a bond ETF, the ingredient is *bonds*. The wrapper affects convenience, trading, taxes, and fees, but it doesn't change what you fundamentally own.

This is liberating. It means you can build a straightforward portfolio without becoming a security analyst. You can choose your ingredients (stocks, bonds, cash) and then choose a sensible wrapper to hold them.

How Stocks and Bonds Can Both Lose Money (And Why That Doesn't Break the System)

One reason people get disillusioned is discovering that "diversification" isn't a magic spell. There are times when both stocks and bonds struggle—most commonly when inflation is rising quickly and interest rates are moving up. Stocks can fall because higher rates make future profits less valuable in today's dollars and because the economy can slow. Bonds can fall because newly issued bonds offer higher yields, pushing down prices of existing bonds.

That's not a failure of asset classes. That's the market repricing a new reality. The correct response usually isn't to throw up your hands and declare everything rigged. It's to zoom out and remember what each piece is supposed to do:

- Stocks are the long-term growth engine.
- Bonds are the stability and income engine (with interest-rate sensitivity).
- Cash is the immediate-liquidity engine.

A portfolio is a system. Systems aren't judged by whether every component wins every year. They're judged by whether the whole machine keeps running when conditions change.

Action That Actually Helps: Label Your Money Before You Invest It

Here's a concrete habit that prevents a lot of future pain: before investing a dollar, give it a job title.

Write down three buckets on paper:

- *Soon money:* needed within a few years (emergency fund, near-term big purchase).
- *Steady money:* meant to support medium-term goals and reduce volatility.
- *Long money:* for goals a decade or more away (often retirement).

Then match the job to the ingredient:

- Soon money mostly wants cash and very high-quality short-term bonds.
- Long money can carry more stocks because it has time to heal after downturns.

Do that, and the market stops feeling like a casino because you stop walking into the wrong room with the wrong expectations. The gas pedal goes where speed is useful. The brakes go where control is necessary. And the money you can't afford to lose next year never gets asked to survive a storm it wasn't built for.

4.14 The Power of Diversification

Most investing mistakes don't start with greed. They start with love.

Someone loves their employer and buys a pile of company stock in the 401(k). Someone loves a product and decides the stock is "obviously" going up. Someone loves a story—*this* new technology, *this* charismatic CEO, *this* once-in-a-generation opportunity—and concentrates their savings into one shining idea.

Concentration feels courageous. Diversification feels like ordering the plain oatmeal.

Then reality shows up with its one consistent personality trait: it's allowed to surprise you.

Diversification is the financial equivalent of building your house with more than one support beam. It won't make the wind stop blowing. It changes whether one bad gust becomes a collapse.

Two Kinds of Risk: The Storm and the Loose Floorboard

There are two broad kinds of investing risk, and mixing them up is how people end up with portfolios that look "aggressive" but are actually fragile.

Market risk is the storm risk. Recessions, inflation shocks, interest-rate changes, wars, pandemics-events that can push *most* stocks down at the same time. You can't diversify market risk away completely if you own risky assets. If you want the return potential of stocks, you have to accept that sometimes the whole neighborhood floods.

Idiosyncratic risk is the loose-floorboard risk. It's the risk that *one specific company* messes up: fraud, a product failure, a lawsuit, a management disaster, a competitor eating their lunch, or simply a slow slide into irrelevance. Idiosyncratic risk is not rewarded in the long run the way market risk can be, because it's avoidable. You're not being paid extra to bet your future on whether one particular executive team will make fewer mistakes than the rest of humanity.

That's why diversification is often called the only "free lunch" in finance. You can reduce idiosyncratic risk dramatically without necessarily reducing expected return-simply by owning *more* independent sources of return instead of tethering your outcome to one.

The casino version of investing is betting on *a winner*. The grown-up version is owning *the game*.

A Concrete Gut-Punch Example: Enron and the Double-Risk Trap

Enron wasn't just a stock that collapsed. It was a masterclass in how concentration can break a family's finances in two different ways at once.

In the late 1990s and early 2000s, many Enron employees held significant amounts of Enron stock in their retirement accounts-often because of employer stock programs and the very human belief that "we know this company; we work here." When the company imploded in 2001, the stock price fell from extraordinary heights to near zero. For employees who were concentrated, the damage wasn't only "my portfolio is down."

It was also: *my job is gone at the same time my savings are gone.*

That's concentration's worst trick: it can align your life risks. If your income and your investments depend on the same company, one corporate failure can hit your paycheck and your net worth in the same month. Diversification isn't just about owning many stocks. It's about not building your financial life so that one domino knocks over all the others.

A diversified investor can still suffer when markets fall. But they're less likely to suffer the uniquely brutal experience of learning that their "safe" employer stock was actually a single point of failure.

What Diversification Actually Does (and What It Doesn't)

Diversification isn't a guarantee that you won't lose money. It's a guarantee that you won't lose money for a dumb reason.

A dumb reason is: one company blew up and took your future with it.

A smart reason-if we can use that word for painful things-is: the whole economy went through a rough period, and risky assets temporarily fell in value. That kind of loss is the price of admission for growth. It's not fun, but it's at least a price that comes with a reward on the other side: participation in broad business recovery over time.

Here's what diversification *does*:

- It reduces the damage from a single company's failure.
- It smooths the ride by spreading exposure across different industries and business models.
- It increases the odds that some holdings will be doing okay when others are struggling.

Here's what diversification *does not* do:

- It does not stop the entire market from falling.
- It does not protect you from every scenario (especially short-term needs and forced selling).
- It does not replace a sensible asset mix (stocks vs. bonds vs. cash).

A diversified stock portfolio can still drop 40% in a major bear market. The point is that it's unlikely to drop 100% because one CEO committed fraud or one product line got disrupted.

The "Safety Net" Idea: Why a Basket Beats a Spear

Think of a concentrated portfolio like walking a tightrope with a spear: impressive, precise, terrifying. A diversified portfolio is like walking with a net below you. You can still fall. You're just far less likely to be destroyed by one misstep.

The difference is mathematical, not mystical. If you own one stock and it goes to zero, your portfolio goes to zero. If you own 500 stocks and one goes to zero, the portfolio barely notices—assuming the positions are reasonably balanced.

That's not just comfort. It's freedom. It means you can be wrong about a company and still be fine. You can miss the "next big thing" and still build wealth. You can stop chasing perfect predictions and start building a system that survives imperfect reality.

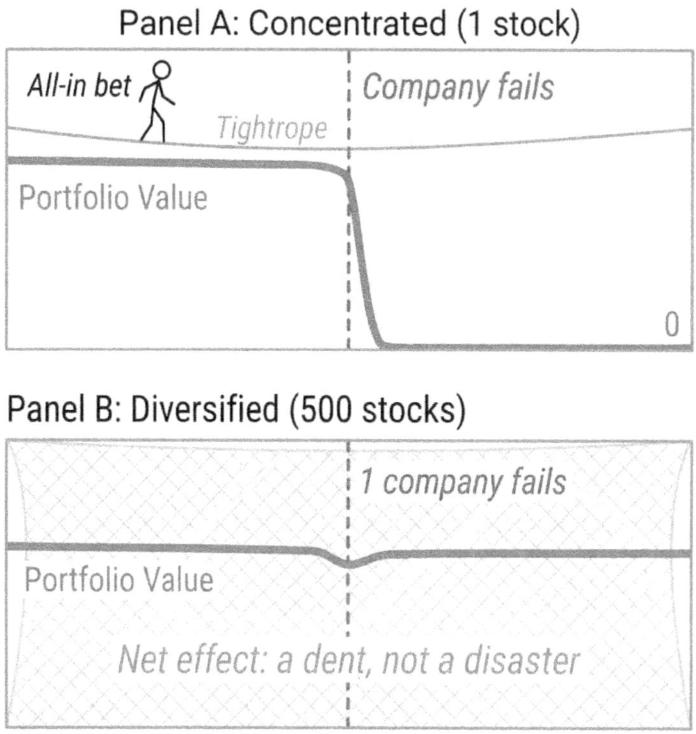

Diversification Isn't "More Stuff." It's "Different Stuff."

A portfolio with 30 stocks can still be dangerously concentrated if they're all the same kind of company. If you own 30 tech companies, you don't have 30 independent bets-you have one big bet that tech will keep being loved by the market at the same time.

Real diversification is about different *drivers*. Different ways companies make money. Different industries. Different countries. Different sizes of companies. Sometimes, different asset classes.

You'll see this in the real world when a particular sector gets a cold. Oil prices drop and energy companies get hit. Interest rates jump and certain fast-growing companies get repriced. A new regulation lands and one industry is suddenly living under a different rulebook. If your entire portfolio is built from one theme, you're not diversified-you're themed.

Themes can be profitable. They can also be expensive hobbies.

Correlation: Why Two Things That Look Different Can Still Fall Together

There's a finance word that matters here: *correlation*. In plain language, correlation is how often two investments move in the same direction at the same time.

When correlation is high, you can own many things and still get the same result as owning one thing: everything rises together, and everything falls together. That's why "diversifying" by buying five different cryptocurrencies, or five different companies in one industry, often fails the moment you need it most.

Low correlation is the magic ingredient. It's what makes a portfolio feel less like a roller coaster. It's also why mixing stocks with high-quality bonds and keeping cash for near-term needs can make investing survivable for normal humans with normal jobs and normal stress.

Correlation isn't stable. In severe market panics, correlations often rise—meaning many risky assets drop together. That's not a reason to give up. It's a reason to understand what diversification can realistically promise: fewer single-company disasters, and a smoother ride most of the time—not immunity from every storm.

The Diversification Trap: When "Safe" Means "Familiar"

The most common concentration risk isn't a flashy meme stock. It's something that feels responsible: the familiar blue-chip company, the employer stock, the "I've heard of it, so it must be safe" portfolio.

Familiarity is comforting, but it's not a risk-control strategy. In fact, familiarity can be a risk amplifier because it makes you overconfident. You stop asking hard questions. You stop noticing how much of your net worth is tied to one name.

A simple stress test helps: if one company in your portfolio dropped 60% next month, would your plan still work? If the answer is "I'd be wrecked," you're not investing-you're balancing your future on one leg.

How Diversification Shows Up in a Normal Person's Portfolio

For most people building wealth from zero, diversification doesn't look like a spreadsheet with 47 holdings. It looks like choosing broad funds that already contain many companies.

A total stock market fund can own thousands of stocks. A global stock fund can spread exposure across countries. A bond fund can hold many bonds so that one issuer default isn't the whole story.

This matters because diversification has two layers:

- *Within an asset class:* owning many stocks instead of one.
- *Across asset classes:* mixing stocks with bonds and cash so you're not betting your short-term life needs on long-term volatility.

You don't need complexity. You need coverage.

Rebalancing: The Boring Move That Quietly Buys Low and Sells High

Diversification is a decision you make once. Rebalancing is how you keep that decision from drifting into something else.

If stocks soar for a few years, a portfolio that started as "moderate" can quietly become "aggressive" because stocks grow to dominate the mix. Rebalancing means trimming what has grown too large and adding to what has lagged, bringing your portfolio back to its intended proportions.

This sounds like a technical chore. It's actually a behavioral hack. It forces you to do the hardest thing in investing in a rule-based way:

- sell a little of what feels great (the recent winner),
- buy a little of what feels disappointing (the recent loser).

People say "buy low, sell high" as if it's a motivational poster. Rebalancing turns it into a habit.

Rebalancing can be done on a schedule (say, once or twice a year) or when allocations drift beyond a threshold (for example, if your stock allocation moves more than a few percentage points away from target). The specific rule matters less than having a rule you can follow when your emotions are loud.

A Practical Takeaway You Can Apply Immediately

Open your retirement account or brokerage account and do one quick, clarifying audit:

- Identify your *largest single holding*. If it's more than 10% of your investments, write down why you're comfortable with that risk.
- Identify whether your income is tied to that holding (employer stock or same industry). If yes, treat that as an extra reason to diversify.
- If you're using broad funds, check whether they overlap heavily (multiple funds that all own the same biggest tech names can look diversified but behave like one bet).

Then choose one small action that improves resilience without requiring a personality transplant: route new contributions toward a broad fund you don't already own, trim employer stock gradually if it's become a dominant slice, or set a calendar reminder for a simple annual rebalance. Diversification works best when it's not a dramatic makeover—just a steady habit of refusing to let one story control your outcome.

4.15 The High Cost of Fees

The most dangerous expense in investing doesn't feel like an expense. It doesn't show up as a big red "charge" in your bank account. It doesn't trigger the pain receptors the way paying $4.99 for shipping does. It just quietly skims a little off the top, day after day, year after year, while you're busy living your life.

Investment fees are the financial equivalent of a slow leak in a tire. You can drive for a long time without noticing. Then one day you wonder why you're working so hard and getting nowhere near as far as you expected.

The brutal part: fees don't have to be outrageous to do real damage. A 1% annual fee sounds polite. It even sounds small-smaller than the sales tax on a sandwich. But investing is the one place where "small" can become "huge," because the fee doesn't just reduce your money *today*. It reduces the base that would have compounded *tomorrow*.

The Reverse Magic Trick: Compounding Works Against You Too

Compounding is usually marketed like a miracle: earn returns on your returns, and time does the heavy lifting. Fees use the same math with a different mission. They compound in reverse.

Here's why the pain is so invisible. If a fund charges 1% per year, you don't write a check for 1% of your account balance. The fund takes its slice internally, so your returns simply show up a little smaller than they otherwise would. It feels like "the market was a bit weak" or "this fund just didn't do great," when in reality part of the underperformance is pre-baked.

And unlike a one-time commission, an annual fee charges you *more* as your account grows. The better you do, the more you pay. It's like agreeing to a lifetime contract where the price increases specifically because the product works.

The Fee You See vs. the Fee You Pay

Investors often fixate on the fees that are easiest to spot and ignore the ones that matter most.

The fees you see are things like trading commissions or a visible account charge. Many brokerages have reduced these, which is good-just not the whole story.

The fees you pay are often embedded:

- *Expense ratio:* the annual operating cost of a fund, expressed as a percentage (for example, 0.05% or 1.00%). It's taken out of the fund's assets, which means it quietly reduces your return.
- *Sales loads:* a commission-like charge to buy or sell certain mutual funds (less common in modern workplace plans, still present in some places).
- *Plan-level 401(k) fees:* administrative fees, recordkeeping, and other costs that may show up explicitly or be bundled into investment options.
- *Trading frictions:* bid/ask spreads and market impact costs-harder to see, but real.

The villain of everyday investing is usually the expense ratio. It's the easiest to compare, the easiest to control, and the one that most reliably drags long-term results.

A One-Percent Difference That Becomes a Life Difference

Suppose two people invest the same way: same contributions, same market returns, same discipline, same luck. The only difference is cost.

One pays a low all-in fund cost. The other pays a fee that's 1% higher every year.

That sounds like a rounding error until you watch what happens over decades. A fee is not a haircut. It's a shave taken every single morning from your future.

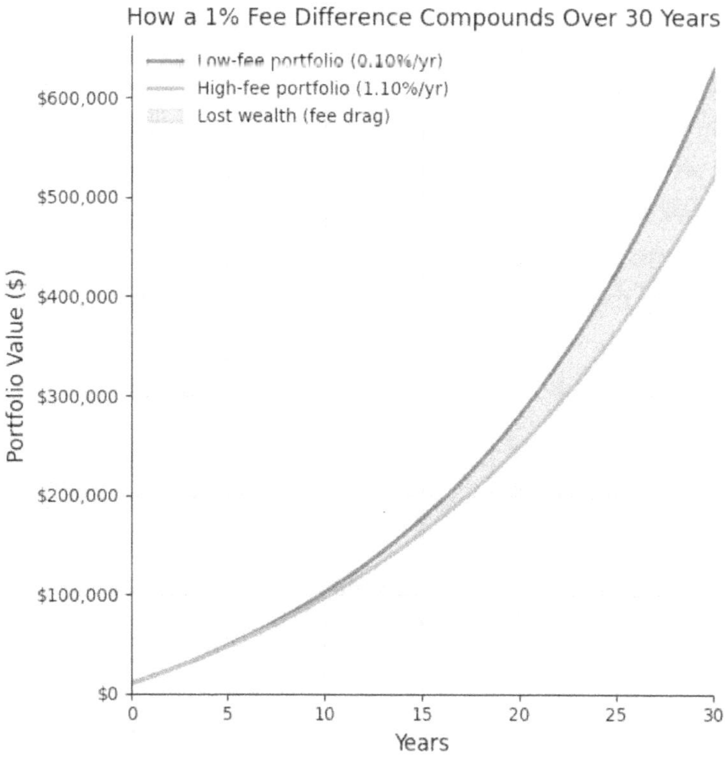

The exact numbers will vary with returns and contributions, but the shape of the outcome is stubborn: two lines start close and then peel apart like a zipper. That red gap isn't bad investing behavior. It isn't a market crash. It's the price of a product you may not even remember choosing.

And this is why fee conversations can feel emotional. They imply something uncomfortable: you can do everything "right" and still fall short simply because you paid too much for the wrapper.

A Real-World Case Study: The 401(k) You Didn't Choose (But You're Paying For)

Many people don't pick their first investing platform. Their employer does. You show up to the job, you get a benefits packet, and there's a retirement plan with a list of funds. You're told to "pick something." Most people pick based on the fund names, past performance, or whatever looks familiar. Fees are either hidden in a PDF, buried in a prospectus, or displayed in a way that looks too small to matter.

Here's a common scenario:

Two employees at two companies, same salary, same saving rate. One company offers a 401(k) with a low-cost index fund charging 0.03% per year. The other offers mostly actively managed funds charging around 1.00% per year, plus a plan administrative fee.

Ten years in, both employees feel like they've been doing the responsible thing: contributing, taking the match, ignoring the noise. But one employee's account grows with less friction, like a runner on smooth pavement. The other is running in sand. Not because of laziness. Because of costs.

This is where the Department of Labor's framing is useful: 401(k) fees generally fall into plan administration fees, investment fees, and individual service fees. You don't need to memorize those categories. You just need to recognize that the biggest recurring drag is often the investment fee-the expense ratio inside the funds you're offered. In other words, the menu matters.

Even if your employer plan has limited options, knowing the fee landscape helps you make the best choice *within* the constraints. If the plan match is strong, you may still prioritize contributing enough to get it-because a match can overwhelm fees. But beyond that, you might choose cheaper funds, or use an IRA for additional savings if it offers lower-cost choices.

Fees aren't just a "Wall Street problem." They're a workplace problem. And the fix is surprisingly unglamorous: reading the fine print like you're protecting your future self—because you are.

``But the Fund Is Good": The Seductive Lie of Gross Performance

High-fee funds often advertise performance before fees or highlight a strong recent streak. That's not always deception; it's marketing. But it creates a mental trap: you start evaluating investments like a sports fan. Who's hot right now? Who won last year?

Fees force a different question, the kind that makes adults unpopular at parties: *After all costs, what are the odds this option beats the cheaper option?*

A higher-cost fund has to outperform a lower-cost fund just to *tie* after fees. That's a steep hurdle, and it gets steeper when you add trading costs inside the fund and taxes in taxable accounts. Skill exists in the world, but the arithmetic is unforgiving: cost is a headwind that never takes a day off.

This is why cost control is so powerful. You can't control whether the market has a great decade or a lousy one. You can control whether you're voluntarily handing away 1% per year for 30 years.

The Sneakiest Fee Is the One You Don't Recognize as a Fee

Not every "fee" is labeled fee.

Sometimes the cost shows up as:

- *A fund that trades a lot.* Higher turnover can mean higher trading costs and, in taxable accounts, more taxable distributions.
- *A complicated product with extra layers.* If a product is hard to explain, it can be hard to price-and complexity often carries a cost.
- *A premium for convenience you don't need.* Some services are worth paying for. Many are just expensive versions of something simple.

This doesn't mean you should never pay for advice or management. It means you should insist on clarity. If you can't explain what you're paying and what you're getting, the odds are not in your favor.

How to Read a Fund Fee Like a Pro (In Under Two Minutes)

You don't need to become an expert. You just need a short checklist that prevents expensive defaults.

When you look at a fund option, find:

- *Expense ratio* (sometimes called "net expense ratio" or "total annual fund operating expenses").

- Any *sales load* (front-end or back-end) if it's a mutual fund.
- Any *additional plan fee* if you're in a 401(k).

Then translate the percentage into dollars. Percentages are soothing; dollars are honest.

If you have $50,000 in a fund with a 1.00% expense ratio, that's about $500 per year. Every year. And if the account grows, that annual number grows. At $200,000, it's $2,000 per year. That's not "small." That's a car payment.

When Paying More Can Make Sense (and When It's Just Charity)

Sometimes higher fees can be worth it:

- You need comprehensive, fiduciary financial planning and behavior coaching, and you'll actually follow through because someone is guiding you.
- You have a complex situation (business ownership, significant tax planning needs, estate planning coordination) where specialized advice is valuable.

What doesn't make sense is paying high ongoing fees for plain-vanilla exposure you can get cheaply-like broad U.S. stock market exposure-without a clear, believable reason that the extra cost will deliver extra value *net of fees*. "It sounds sophisticated" is not a reason. "It had a great five-year chart" is not a reason. "My coworker said it's good" is definitely not a reason.

If a product can't explain its value in plain language, assume the plain-language translation is: it's expensive.

The Actionable Move: Run a Fee Audit on Your Biggest Accounts

Pick your largest investing account (401(k), IRA, or brokerage). Then do this:

1. List each fund you own and its expense ratio.
2. Circle anything above 0.50%. Put a star next to anything above 1.00%.
3. For each circled fund, ask one question: *Can I get similar exposure with a lower-cost fund in this account?*
4. If you can, change future contributions first (easy, low drama), then consider exchanging existing holdings (mindful of taxes in taxable accounts).

This is not about perfection. It's about refusing to bleed quietly.

There's a particular kind of confidence that comes from knowing your investing plan isn't secretly feeding a fee machine. You still can't control markets. You still can't control headlines. But you can control whether your returns get siphoned off in the fine print—and that alone can be the difference between "I invested for decades" and "I actually got paid for it."

4.16 The Case for Passive Investing

There's a particular kind of pain that only investing can deliver: you make a reasonable choice, you do the research, you pick the "better" fund—and then you watch a boring, no-personality index fund beat you anyway. It feels unfair, like showing up to a cooking contest with hand-ground spices and losing to someone who microwaved oatmeal

That frustration is also a clue. In markets, the hardest opponent isn't another investor. It's the math.

Passive investing is not a claim that people are stupid, or that analysis is useless, or that markets are always right. It's a practical decision to stop volunteering for a game where the average player loses after costs. The passive move is humble and, for most households, brutally effective: buy a diversified slice of the market, keep fees low, rebalance occasionally, and let time do what time does.

Active vs. Passive: Two Styles, Same Ingredients

Active and passive are not asset classes. They're styles of implementation applied to the same building blocks.

Active investing means someone is making bets about what to own, when to buy, and when to sell—trying to outperform a benchmark (like the S&P 500). That "someone" might be you, a fund manager, or an algorithm.

Passive investing means accepting the benchmark as the goal. You "buy the haystack" instead of hunting for needles: own an index fund that tracks a broad market, minimize trading, and minimize costs.

Both approaches can own stocks and bonds. The difference is whether you're trying to be smarter than the crowd, or whether you're trying to capture what the crowd (the entire market) earns.

The Core Reason Passive Wins: The Arithmetic Is Rigged (By Reality, Not by Conspiracy)

Here's the least glamorous but most important insight in investing: *before costs, active investors as a group are the market*. They *are* the market because the market is just the collection of everyone's holdings. Every share of stock is owned by someone. Add up all active portfolios, and you get the market portfolio.

So, before fees and trading costs, the average active dollar earns the market return. That's not an opinion. It's a definition.

Now subtract reality:

- management fees,
- transaction costs,
- taxes triggered by trading (especially in taxable accounts),
- the bid/ask spread and other frictions.

After those costs, the average active dollar must underperform the market. This logic is sometimes called the arithmetic of active management, and it's as close as investing gets to a law of physics.

This is why passive investing feels almost "too easy." It's not easy because markets are easy. It's easy because you're choosing the side of the equation that doesn't require winning a zero-sum competition after paying a toll.

The Problem Isn't That Active Managers Are Bad. It's That They're Expensive Competitors

It's tempting to frame this as "professionals are dumb." That's not true, and it's not the point.

Many active managers are smart, hardworking, and deeply informed. The problem is that they're competing against:

- other smart, hardworking, deeply informed managers,
- institutions with faster data and better execution,
- teams of PhDs optimizing strategies,
- investors with different constraints (some are forced buyers or forced sellers).

If you imagine this as a sport, it's not your neighbor's pickup basketball game. It's the NBA. And then you have to pay the league a fee just to step onto the court.

Even if skill exists, the hurdle is high: an active approach has to overcome its own costs and still beat the benchmark. That's why "some managers outperform" can be true while "most investors should be passive" is also true. The existence of a few exceptional chefs doesn't mean most people should pay restaurant prices every night instead of learning a simple, healthy meal at home.

A Real-World Case Study: The Graveyard of Yesterday's Stars

Every few years, a fund manager becomes a celebrity. Their fund shows incredible performance, money pours in, media profiles bloom, and the story writes itself: genius has been discovered.

Then something predictable happens. Performance cools. The market environment changes. The strategy stops fitting the moment. The fund grows so large it can't move as nimbly. Or the manager takes more risk to keep the streak alive. The headlines fade, the inflows slow, and investors who arrived late are left holding the bill.

This pattern shows up again and again because of a simple human flaw: we chase what worked recently, and we confuse a hot streak with a permanent advantage. In investing, recent outperformance is often less like "proof of skill" and more like "proof of what the market rewarded during that specific period."

You can see the same dynamic at the individual stock level. Investors pile into last year's winners because it feels safe to follow the crowd. But the crowd is often late. By the time an investing idea is obvious and everywhere, it's frequently already priced in.

Passive investing dodges this trap by refusing to chase. You still own the winners—but only in proportion to their place in the market, not because you fell in love with a narrative at the top.

The Three Hurdles Active Investing Must Clear (and Usually Doesn't)

For an active approach to beat a simple index fund over time, it has to clear multiple hurdles simultaneously:

Hurdle 1: Pick better holdings than the market. Not just "good companies," but companies that will perform better than the market expects. The market already knows about great companies. Outperformance requires being right when the consensus is wrong.

Hurdle 2: Time it well enough. Even if you pick a future winner, you can still lose if you buy after hype inflates the price or sell in panic during a temporary drawdown.

Hurdle 3: Do all of this after costs. The index fund is standing there with a low expense ratio, minimal trading, and no ego. It's hard to beat a competitor that charges almost nothing and makes almost no unforced errors.

This doesn't mean active investing is impossible. It means it's not the default choice for someone whose main goal is building wealth steadily while also living a real life.

Passive Investing Is a Behavioral Strategy Disguised as a Market Strategy

Most investing underperformance isn't caused by picking the "wrong" asset class. It's caused by behavior: buying high, selling low, and changing strategies mid-storm.

Passive investing helps because it reduces the number of high-drama decisions you have to make. It narrows your job to a few repeatable actions:

- choose broad, diversified funds,
- keep costs low,
- keep contributing,

- rebalance when needed.

That simplicity is not a lack of sophistication. It's protection against a predictable human pattern: we do the worst possible thing at the worst possible time when we feel threatened.

A passive investor can still panic—but the system gives them fewer levers to pull. Fewer levers means fewer ways to accidentally blow up a long-term plan on a bad Tuesday.

``But What About the Big Opportunities?"

This is the emotional objection: *If I index, don't I miss the next Apple? The next Amazon? The next breakthrough?*

No. You own them automatically as they become meaningful parts of the market. You may not own them at the absolute earliest moment, when the risk is highest and the outcome is most uncertain. But you also avoid the graveyard of "next big things" that never arrive.

Indexing is a commitment to reality: you're not trying to predict which companies will dominate the future. You're betting that *some* companies will, and you'd like to own the winners when they emerge—without having to guess which seedlings will become trees.

There's a deep peace in that. You stop arguing with the future and start owning it broadly.

What Passive Looks Like in Practice (Without Turning Your Life Into a Spreadsheet)

A practical passive approach usually means using low-cost index funds or ETFs that cover large, diversified markets. Examples of broad exposures people commonly use:

- total U.S. stock market,
- total international stock market,
- broad U.S. bond market (or high-quality bond funds),
- cash equivalents for near-term needs.

The exact mix depends on your time horizon and risk tolerance, but the philosophy stays the same: own a wide swath of productive assets, pay as little as reasonably possible to hold them, and avoid the temptation to "do something" every time the market becomes interesting.

Indexing is not laziness. It's selecting a game you can win.

A Place Where Active Can Make Sense (If You're Honest About the Goal)

Some people enjoy active investing the way others enjoy poker. They like research, they like the challenge, they like the feeling of being right. If that's you, active investing can be fine *as entertainment*, as long as you treat it like entertainment:

- keep it to a small, defined slice of your portfolio,
- accept that the most likely outcome is underperforming after costs,
- avoid letting it interfere with the core plan.

The danger is when "fun money" quietly becomes "future rent money." The market has no sympathy for the difference between a hobby and a necessity.

Another area where active can matter is in certain less-efficient markets or specialized strategies—but those are advanced topics, and even then, the right question is still the same: *Is the expected benefit worth the cost, complexity, and behavioral risk?*

The Actionable Move: Pick a Benchmark and Stop Fighting It

Here's a simple step you can take immediately: choose a benchmark that matches your goal, then align your investing behavior to it.

If your goal is long-term growth for retirement, a broad stock index benchmark is a sensible reference point. If your goal includes stability, a blended benchmark (stocks plus bonds) may be more appropriate. The benchmark isn't a report card for your ego. It's a steering wheel. It tells you what "success" looks like so you're not constantly tempted to chase whatever is currently getting attention.

Then make one concrete change:

- If you're in high-fee actively managed funds, redirect new contributions to a low-cost index option in the same account.
- If you're stock-picking with your core savings, decide what percentage is your "hobby sleeve" and cap it.
- If you keep changing strategies, write down a simple rule: *I only rebalance on a schedule or when allocations drift beyond a set range.*

Passive investing wins the way a well-built house wins: not by being exciting, but by still standing after decades of weather. You don't need to predict which way the wind blows next week. You need a structure that turns time—messy, unpredictable time—into something you can actually keep.

4.17 Choosing the Right Investment Vehicle

If asset classes are the ingredients, the investment vehicle is the container you store them in. The ingredient determines what it *is* (stocks, bonds, cash). The container determines how it behaves in real life: how it's taxed, how easily you can move it, how much it costs to hold, and what rules you must follow to access it.

People often reverse this. They spend hours debating ETFs versus mutual funds, then treat taxes and account type like an afterthought. That's like arguing about whether to use a glass jar or a plastic one while ignoring the fact that one pantry is refrigerated and the other is sitting in the sun.

There are two big "container" decisions that matter most:

- *Account type:* where you hold investments (401(k), 403(b), IRA, Roth IRA, taxable brokerage).
- *Fund wrapper:* how you buy diversified baskets (ETF vs. mutual fund, index fund vs. active fund).

Nail those, and the rest is mostly detail.

First Choose the Account: The Tax Rules Are the Real Game Board

A great investment held in the wrong account can underperform a decent investment held in the right one, simply because taxes and rules change your net result.

Workplace plans (401(k), 403(b), etc.): the "easy mode" you didn't know you had

A workplace retirement plan is often the most powerful on-ramp for building wealth because it's designed to make good behavior automatic:

- contributions can come straight from payroll,
- money can be invested without you remembering every month,
- many employers offer a match (free money with rules).

If your employer offers a match, that match is usually the highest "return" you can get with near-zero risk. Skipping it is like refusing a raise because the paperwork is annoying.

The trade-off is control. Your plan menu may be limited. Fees may vary. There may be plan-level administrative costs. Still, for most people, the match and the tax advantages are so valuable that it's worth starting here.

A practical, non-glamorous move: look at your plan's fund list and identify the lowest-cost broad index options. You don't need a perfect menu to make a strong plan. You need a couple of low-fee building blocks.

IRAs: your personal retirement account, your rules (mostly)

An IRA (Individual Retirement Account) gives you more control than most workplace plans. You choose the provider and the investment options, which often means you can choose low-cost index funds easily.

For tax purposes, the big fork is *Traditional* vs. *Roth*:

- *Traditional IRA:* contributions may be tax-deductible depending on income and workplace plan coverage; growth is tax-deferred; withdrawals in retirement are generally taxed as income.
- *Roth IRA:* contributions are made with after-tax dollars; growth can be tax-free; qualified withdrawals in retirement can be tax-free.

This can sound like a philosophical debate about the future tax code. It's really a bet on your personal story: do you expect your tax rate to be higher later, or lower? Many people split the difference by using both types over time when possible.

A concrete 2026 anchor: the IRA contribution limit for 2026 is $7,500 (or $8,600 if you're age 50 or older, thanks to catch-up contributions). Roth IRA eligibility phases out at higher incomes; for 2026, phaseouts rise (for example, the married filing jointly phaseout begins at $242,000 of modified adjusted gross income, and the single phaseout begins at $153,000). These numbers matter because they determine whether "just open a Roth" is an option or whether you need a different route.

Taxable brokerage accounts: maximum flexibility, minimum tax shelter

A taxable brokerage account is the investing account with no retirement label. That's its strength and its weakness.

Strength: you can use the money anytime without retirement-account withdrawal rules. This makes taxable accounts great for goals that aren't retirement: early financial independence, a future home purchase far enough away to tolerate volatility, or simply building wealth with fewer constraints.

Weakness: you're exposed to taxes along the way:

- dividends can be taxable each year,
- selling at a gain triggers capital gains tax,
- high-turnover funds can generate more taxable distributions.

This is where low-cost, low-turnover index funds and tax-aware strategies can really shine. In a taxable account, the "best" fund is not just the one with the highest pre-tax return. It's the one with the best *after-tax* outcome.

A Real-World Scenario: Three Accounts, One Person, One Smart System

Meet Priya. She's 29, early in her career, and finally has room in her budget to invest $500 per month. Her company offers a 401(k) with a 4% match. She also has student loans and wants the option to buy a home in maybe 7–10 years.

Priya's choices aren't about finding the hottest ETF. They're about assigning dollars to containers based on rules and goals:

- She contributes enough to her 401(k) to get the full match. That's free money plus tax advantages.
- She opens a Roth IRA and invests additional retirement savings there for tax-free growth potential and flexibility.
- If she still has extra after that, she uses a taxable brokerage for her longer-term, non-retirement goals—but she keeps her near-term home down payment money out of stocks because she doesn't want a market downturn to dictate her life timeline.

This is what "choosing the right vehicle" looks like in the real world: the same basic ingredients, placed into containers that match the job.

Then Choose the Wrapper: ETF vs. Mutual Fund

Once you've picked the account, you're usually choosing between two common ways to hold diversified investments: ETFs and mutual funds. Most of the time, they can own extremely similar portfolios. The difference is how you buy and sell them, and how the mechanics affect taxes and convenience.

Mutual funds: simple, steady, and automation-friendly

A mutual fund typically transacts once per day at its *net asset value* (NAV), which is the value of the underlying holdings at the end of the trading day. If you place an order during the day, the trade usually executes after the market closes at that day's NAV.

Practical advantages:

- Easy automatic investing: "invest $200 every paycheck" works smoothly.
- Easy to buy in exact dollar amounts (depending on the provider).
- Often a clean fit inside retirement accounts where trading during the day doesn't matter.

Potential downsides:

- Some mutual funds have higher expense ratios (not always, but often).
- In taxable accounts, some mutual funds can distribute taxable capital gains due to internal activity (varies widely by fund structure and management).

ETFs: trade like a stock, often tax-efficient

An ETF trades on an exchange throughout the day, like a stock. That means the price can move intraday and can be slightly above or below the value of the underlying holdings.

Practical advantages:

- Intraday trading and flexibility: you can place limit orders if you care about execution price.

- Often very low expense ratios, especially for broad index ETFs.
- In many cases, strong tax efficiency in taxable accounts due to the way ETFs handle creations/redemptions (a structural advantage that can reduce taxable capital gains distributions).

Potential downsides:

- You may pay the bid/ask spread (a small implicit cost, usually tiny for large ETFs).
- Automatic investing in exact dollar amounts depends on the brokerage and whether fractional shares are supported.
- The ability to trade all day can tempt you to *act* all day—which is rarely helpful.

A simple way to choose: if you want set-it-and-forget-it automation and you're investing inside a retirement account, a low-cost mutual fund can be perfect. If you're investing in a taxable brokerage and you want low costs and tax efficiency, broad index ETFs are often a strong fit. Either can work; the bigger mistake is overpaying for the exposure you want.

Index Funds vs. Actively Managed Funds: Don't Confuse the Wrapper With the Strategy

An ETF can be passive or active. A mutual fund can be passive or active. The wrapper is the container; the strategy is what's inside.

If your goal is to capture market returns with low friction, you're usually looking for:

- broad diversification (not a narrow theme),
- low expense ratios,
- low turnover (especially in taxable accounts),
- a clear benchmark and a clear mandate.

If you choose active management, do it with your eyes open: higher fees, potentially higher turnover, and the very real possibility of underperforming a simple index over long periods. Active can still be a rational choice for some investors, but it shouldn't happen by accident because a fund name sounded impressive.

The Checklist That Saves You Thousands: Cost, Taxes, Convenience, Behavior

When deciding between vehicles, most people overweight what's easy to see (ticker symbols) and underweight what actually changes outcomes. A better lens is:

Cost. Expense ratios, plan fees, advisory fees. Small numbers that compound into large consequences.

Taxes. Taxable vs. tax-advantaged accounts; how much turnover and distributions a fund tends to generate; whether you're likely to sell before long-term capital gains rates apply.

Convenience. If a vehicle makes good behavior automatic—paycheck contributions, auto-investing, simple rebalancing—it's doing real work on your behalf.

Behavior. The best vehicle is the one that keeps you from sabotaging yourself. If intraday trading will turn you into a compulsive checker, the "flexible" option can become the expensive option, even if the fee is low.

This is where investing gets refreshingly human. Your vehicle choice is partly finance, partly self-knowledge.

A Practical Priority Order (Without Overcomplicating It)

If you want a simple decision order that works for many people:

- Get the employer match in your workplace plan if it's available.
- Use an IRA (Roth or Traditional depending on eligibility and tax situation) for additional retirement savings, especially if it offers lower-cost options.
- Use a taxable brokerage for goals that need flexibility or for savings beyond retirement limits—but be intentional about taxes and time horizon.

This isn't a universal law. It's a default path that tends to align incentives: free match dollars first, tax shelter next, flexibility last.

Actionable Takeaway: Pick One Vehicle and Make It Automatic

The best investment vehicle is the one you actually use consistently.

Choose one account you will fund automatically starting this month:

- If you have a 401(k) match and you're not getting it, set your contribution to capture the full match.
- If you already get the match, set up an automatic monthly transfer into an IRA or a taxable brokerage (depending on your goal and eligibility).

Then choose one low-cost, broadly diversified fund inside that account and set the contribution to buy it on autopilot. A simple, automatic vehicle beats a sophisticated plan you keep meaning to implement. Consistency is a container too: it holds your future together when motivation is unreliable.

Chapter 5

Constructing Your Wealth Portfolio

You have the blueprints; now it is time to pour the concrete. Many people get stuck here, paralyzed by the fear of choosing the wrong account or pressing the 'buy' button at the wrong time. In this chapter, we strip away the complexity of Wall Street jargon. We will build your portfolio layer by layer, starting with the foundation of asset allocation, framing the walls with tax-advantaged accounts, and putting on the roof with automation. By the end, you won't just have a plan; you'll have a running engine.

5.1 The Chef's Special: Designing Your Asset Mix

Imagine two friends at the same restaurant. One orders the chef's tasting menu and happily eats anything put in front of them. The other scans every ingredient like it's a contract, swapping sides, asking for sauce on the side, customizing the entire experience. Investing attracts both types. The first person wants a *system*. The second wants *picks*.

Most new investors assume wealth is built in the customization layer: finding the "best" stock, the hottest sector, the perfect entry point. That's a tempting story because it gives your brain a hero to root for: *the winning pick*. But the quiet truth is that the biggest driver of your investing experience usually isn't which individual stock you chose—it's the *mix* of assets you decided to hold in the first place.

That mix has a name: *asset allocation*. In plain English, asset allocation is simply how you split your money among major categories like stocks, bonds, and cash. It's the recipe, not the garnish. It determines how much your portfolio tends to grow when times are good *and* how much it tends to flinch when the world gets weird. If you get this recipe right, your investing life gets calmer, more predictable, and dramatically easier to stick with.

The easiest way to feel this is to stop thinking like a day trader and start thinking like a chef plating dinner.

Your Portfolio Is a Dinner Plate (Not a Trophy Case)

A trophy case is where you display individual wins. That's how stock picking feels: you want a lineup of brag-worthy names. A dinner plate is different. A plate is about *balance*. It's about getting enough of what helps you grow, and enough of what keeps you steady.

On your investing plate, stocks are the protein: they're the muscle-builder, the long-term growth engine, the thing that tends to do the heavy lifting over decades. Bonds (and sometimes cash) are the vegetables: not as exciting, but they stabilize the meal. They help you avoid the investing version of a sugar crash—panic-selling after a sharp drop because your portfolio is swinging too wildly for your stomach to handle.

There's no universally "correct" plate. A person training for a marathon and a person recovering from surgery eat differently. Same with money. A 25-year-old investing for retirement in 40 years can usually tolerate a spicier plate than someone who plans to use the money for a home down payment in three years. The point isn't to mimic someone else's meal. The point is to build a plate you can actually finish.

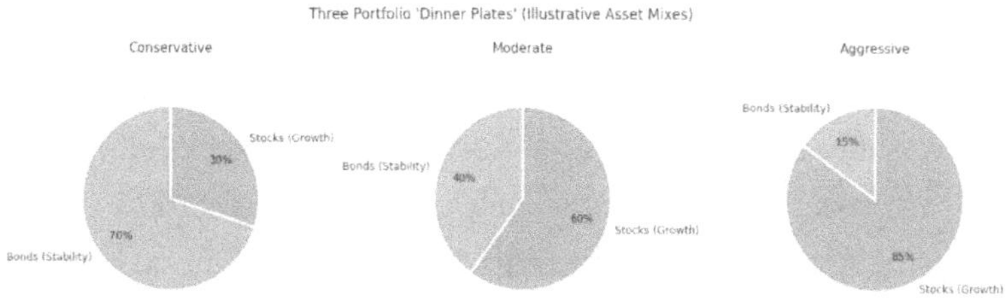

Those three plates are not commandments; they're conversation starters. What matters is the idea hiding in plain sight: if you pick an aggressive plate, you're choosing a life with more volatility. If you pick a conservative plate, you're choosing a life with less volatility but typically less growth potential. This is not a moral decision. It's a physics decision.

Asset Allocation: The Decision That Shows Up on Your Worst Day

If investing only happened on good days, everyone would own stocks forever and never sell. The real test is what your portfolio feels like on a day the market drops hard and your group chat suddenly turns into a disaster movie.

Asset allocation is the decision that determines whether you can keep your hands off the controls when turbulence hits.

Here's the uncomfortable reality: the biggest investing mistakes aren't usually made because someone didn't know the difference between a market order and a limit order. They're made because someone built a portfolio that was too scary for their nervous system. When fear shows up, they hit eject: sell after a crash, wait too long to buy back in, and turn a temporary decline into a permanent setback.

A calmer plate doesn't prevent losses; it helps prevent *you* from making them worse.

There's a reason this matters more than stock picking. Even if you choose the "right" stocks, a portfolio that's 100% stocks will still drop when markets drop. Your picks can be brilliant and still get dragged down by the tide. Meanwhile, someone with an allocation that matches their temperament might hold a perfectly boring set of broad funds—and quietly win simply by staying invested.

Stock Picking vs. Portfolio Building: The Difference Between a Brick and a House

A stock is a brick. A portfolio is the house.

You can buy gorgeous bricks and still build a house that collapses if you ignore the blueprint. Asset allocation is the blueprint: it decides how much of your house is built out of materials that expand and contract with the weather.

Stock picking tries to answer: *Which companies will beat expectations?*

Asset allocation answers: *How much risk am I taking as a lifestyle choice?*

Those are different games. Stock picking is often about being right. Asset allocation is about being *consistent*. It's about selecting a level of risk you can survive long enough for compounding to do its job.

Another way to say it: you don't need to be a better predictor than Wall Street. You need a plan that makes it hard to sabotage yourself.

A Concrete Case Study: Two Investors, One Decade, Very Different Outcomes

Meet Maya and Chris. Same age (30), same income, same savings rate: $500 a month invested automatically. Neither is trying to be clever. Both use low-cost index funds. The only difference is the dinner plate.

Maya chooses an aggressive allocation: 90% stocks, 10% bonds. She believes in growth and doesn't want to "waste time" with bonds.

Chris chooses a moderate allocation: 60% stocks, 40% bonds. He wants growth, but he wants a shock absorber.

Then a nasty market decline hits. Maya watches her portfolio fall fast, because the engine is powerful in both directions. She tells herself she's fine—until the headlines go from "volatile" to "crisis." One night, she opens her account, sees that she's down an amount that feels like months of work, and sells most of her stocks "until things calm down."

Chris is also down, but not as violently. His bonds didn't make him rich; they did something more valuable in that moment: they made the drop *survivable*. He keeps contributing.

He doesn't feel heroic. He feels mildly annoyed. That's the point.

A year later, markets rebound. Maya wants to buy back in, but now prices are higher and she feels foolish. She waits for a "better entry." The better entry never arrives. She slowly drifts back into stocks after missing a chunk of the recovery.

Here's the twist: the difference between Maya and Chris is not intelligence. It's not even discipline in the abstract. It's that Chris chose an allocation that made discipline easier. His plate fit his appetite, so he could stick with it under stress.

This is why asset allocation is such a big deal. A theoretically optimal portfolio that you abandon at the wrong time is an expensive fantasy. A slightly less aggressive portfolio you can hold through chaos can outperform in real life—because you actually *stay in the game.*

What Stocks and Bonds Actually Do (When No One Is Selling You Anything)

Stocks represent ownership in businesses. They're messy, competitive, innovative, sometimes irrational, and over long stretches they've historically rewarded investors for enduring uncertainty. The uncertainty is the admission price.

Bonds are loans. When you buy a bond, you're essentially lending money to a government or a corporation in exchange for interest. Bonds tend to have lower long-term return potential than stocks, but they often behave differently during scary periods. That "different behavior" is the whole point: bonds can reduce the size of a portfolio's swings.

Cash (or cash-like investments) is the fire extinguisher. It doesn't build wealth quickly, but it can prevent forced selling. It buys time. And time is the rarest asset when panic hits.

Notice what's missing: none of these are "good" or "bad." They're ingredients. Your job is to combine them into something you can eat for years.

Risk Tolerance Isn't a Personality Trait (It's a Math-and-Sleep Equation)

People talk about risk tolerance like it's an identity: "I'm conservative," or "I'm aggressive," like it's a favorite color. It's more practical than that.

Risk tolerance is the intersection of:

- how soon you'll need the money,
- how much flexibility you have (can you delay the goal? increase savings?),
- and how well you can sleep when your account balance drops.

Time horizon is the most underrated piece. If you need the money soon, you don't have time to wait for the market to recover from a deep drop. If you don't need the money for decades, short-term drops matter less *mathematically*—even if they still feel dramatic emotionally.

Flexibility matters because it determines how catastrophic a downturn would be. If your goal is optional (retire at 55 instead of 60), you can take more risk than if your goal is fixed (tuition payment due in 18 months).

And then there's sleep. If your portfolio keeps you awake, it's too spicy. This isn't weakness. It's realism. A plan you can follow beats a plan you admire.

The Glide Path: How the Plate Changes as the Deadline Approaches

A glide path is a fancy term for a simple idea: as you get closer to needing the money, you generally reduce risk. You don't do this because you suddenly stop believing in growth. You do it because the penalty for a bad year gets harsher when you're near the finish line.

Think of it like this. When you're far from your destination, taking a scenic route is fine. When your flight boards in 30 minutes, you want the most reliable route available.

For retirement investing, this often means gradually shifting from a stock-heavy plate toward a more balanced plate as retirement approaches. Many target-date funds do this automatically, but you can also do it manually by adjusting your allocation every few years.

A glide path isn't about predicting crashes. It's about respecting deadlines.

Diversification: Don't Bet Your Future on One Story

Asset allocation is the split between major asset classes. Diversification is spreading your bets *within* those classes.

Within stocks, diversification means holding many companies across industries and countries rather than relying on one sector to stay hot forever. Within bonds, it can mean holding many issuers and maturities rather than concentrating in a single type of bond.

The emotional benefit of diversification is underrated. When one part of your portfolio disappoints, another part is often doing "less bad," or even fine. That reduces the temptation to make dramatic changes at exactly the wrong moment.

A diversified portfolio won't protect you from every loss. What it can do is prevent a single blow-up from becoming a financial identity.

A Practical Way to Pick Your First Allocation (Without Overthinking It)

A clean starting point is to choose a plate based on a single question:

If my portfolio dropped 30% in a bad year, could I keep contributing and not sell?

If the honest answer is "no," your portfolio is too stock-heavy *for you*, regardless of what any internet stranger claims is "optimal."

From there:

- If your goal is 15+ years away and your income is stable, you can generally afford more stocks.
- If your goal is under 5 years away, consider a much more conservative mix, because time won't reliably rescue you from a bad sequence of returns.

- If you're unsure, moderate is a respectable default. The best allocation is the one you'll still hold when the market is screaming.

Then make it real by writing it down as a target, like a recipe:

$$\text{Target allocation} = 60\%\ \text{Stocks} + 40\%\ \text{Bonds.}$$

That single line is more powerful than a watchlist full of "promising" stocks, because it becomes a rule you can follow when emotions are loud.

The Hidden Benefit: Asset Allocation Turns Investing Into a Repeatable Process

Here's the part nobody advertises because it's not flashy: once you choose an allocation, investing becomes a series of boring, repeatable actions.

You contribute regularly. You buy according to your target mix. When markets move, your allocation drifts, and you nudge it back toward the recipe. That's it. No constant decision-making. No endless debate about whether a stock is "still a buy." Less temptation to tinker.

That reduction in decisions is not a small thing. Decision fatigue is how smart people end up making dumb moves. A clear allocation is like a pre-commitment device: it protects you from your own worst instincts.

Actionable Takeaway: Build Your Plate in 10 Minutes

Pick one investing goal (retirement is fine). Then do this on paper before you ever click "Buy":

1. Choose a target allocation you can live with in a bad year: Conservative (30/70), Moderate (60/40), or Aggressive (85/15) as a starting template.
2. Write one sentence explaining *why* it fits: time horizon, flexibility, and sleep.
3. Commit to holding that allocation through a full market cycle by setting a simple rule: "I don't change my stock-/bond split unless my goal timeline changes."

A portfolio built this way won't impress anyone at a party. It will do something better: it will keep working while you're busy living your life.

5.2 The Tax Reality Check

Taxes are the background music of your financial life. You don't notice them when the song is soft, but over decades they can change the entire mood of the room. Two people can earn the same market returns and end up in different places simply because one of them kept handing little slices of compounding to the IRS every year, while the other let those slices stay invested.

That's the uncomfortable part about taxes: they don't have to be dramatic to be destructive. A one-time bad trade hurts like a paper cut. A quiet, annual tax drag is more like a slow leak in a tire. You can still drive for a while, but the ride gets rougher, and you burn more gas than you needed to.

A *tax reality check* is less about memorizing rules and more about seeing taxes the way an engineer sees friction. You can't eliminate friction, but you can design around it. And the first design step is understanding two rates that get confused constantly: *marginal* and *effective*.

The Two Tax Rates That Control Your Decisions

Your *marginal tax rate* is the tax rate on your *next* dollar of income—the top slice. If you earn one more dollar, how much of that dollar goes to federal income tax? That's marginal.

Your *effective tax rate* is your average rate—total tax divided by total taxable income. If you zoom out and ask, "Overall, what percentage of my income went to federal income tax?" that's effective.

People fear marginal rates because they sound like they apply to everything: "I'm in the 22% bracket," or "I'm in the 32% bracket," said with the same dread as "my flight is canceled." But tax brackets don't work like a single flat price. They work like a progressive set of tiers.

Which brings us to the most stubborn myth in personal finance: "If I earn more and move into a higher bracket, I'll take home less money."

No. Earning more money does not reduce your take-home pay because you entered a higher bracket. It can reduce certain benefits, credits, or subsidies in some cases (that's a separate, real issue), but the basic bracket math doesn't punish the earlier dollars you already earned. The earlier dollars keep their lower rates.

To make this click in your bones, picture tax brackets as buckets.

Tax Brackets as Buckets (Where the Cheapest Bucket Fills First)

Think of each tax bracket as a bucket with a price tag on it. The first bucket is cheap. The next bucket costs more per gallon. As your income (water) pours in, it fills the cheapest bucket first. Only when that bucket is full does the water spill into the next bucket. And only *the water in the higher bucket* gets charged the higher rate.

That's it. That's the whole mechanism. The confusion happens because people hear "you're in the 24% bracket" and assume the government taxes *all* their income at 24%. In reality, you might have income taxed at 10%, 12%, 22%, and 24% all in the same year.

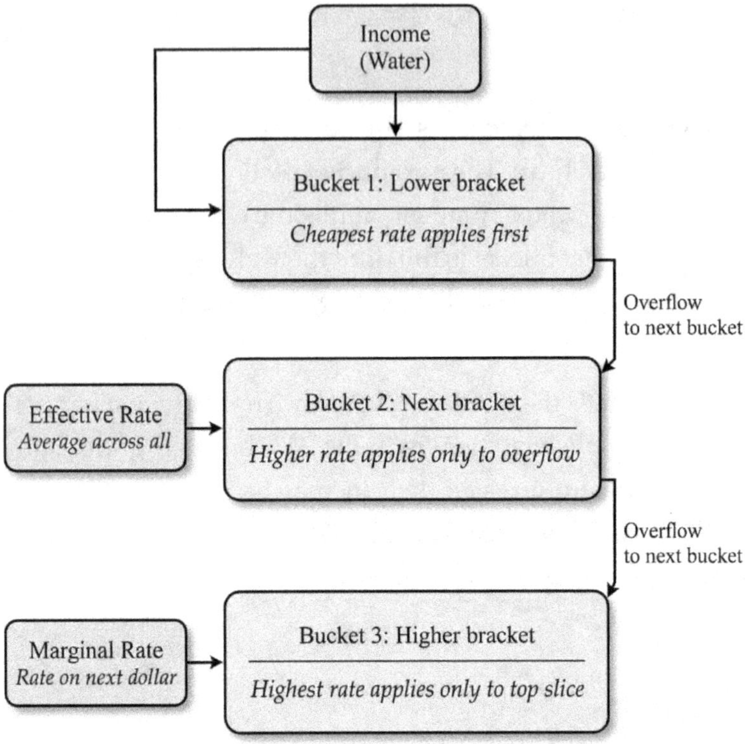

Once you see the buckets, you can stop fearing promotions like they're booby traps. Moving into a higher bracket doesn't retroactively reprice the water in the lower buckets. It just means you finally have enough water to spill into a more expensive bucket.

A Concrete Example: The Raise That ``Put Me in a Higher Bracket"

Suppose Jordan earns $78,000 and gets a raise to $83,000. Jordan hears, "Now you're in a higher bracket," and immediately feels like the raise comes with a hidden fee.

Here's what actually happens in bracket land: only the *top slice* of Jordan's income gets taxed at the higher marginal rate once the lower bucket(s) are filled. The earlier dollars still get taxed at the lower rates they always did.

So the raise still increases take-home pay. Not dollar-for-dollar, because taxes exist—but it increases it.

The real question to ask is not, "Will a raise hurt me?" The real question is, "How much of my next dollar do I keep?" That's marginal rate thinking, and it's useful. It helps you estimate how valuable a deduction is, or whether a pre-tax retirement contribution reduces your taxes meaningfully. But it should never be used as a reason to avoid earning more.

Why Investors Should Care: Taxes Change Your *After-Tax* Return

Investing conversations love to talk about returns as if taxes don't exist. You'll hear: "This fund returned 8%." Great. But what did *you* get to keep?

A simple way to think about it:

$$\text{After-tax return} \approx \text{Pre-tax return} - \text{Tax drag}.$$

Tax drag is the performance you lose to taxes along the way. In a retirement account where growth is tax-deferred or tax-free (depending on the account type), tax drag can be low during the years you're accumulating. In a regular taxable account, tax drag can show up every year through dividends, interest, and realized capital gains.

This is why taxes are a wealth-building topic, not just an April topic. Compounding is powerful precisely because it keeps reinvesting returns. If a portion of those returns gets skimmed off annually, the compounding engine runs on less fuel.

The frustrating part is that the tax drag often doesn't feel like a dramatic event. You don't feel "taxed" when a mutual fund distributes capital gains and you owe money even if you didn't sell anything. You just see a slightly smaller account balance than you expected. That's how the leak stays hidden.

The Four Main Ways Taxes Quietly Nibble at Investments

A practical mental model is to treat your investments like a business with four common "expense lines" created by the tax code:

- *Interest income* (often taxed like ordinary income): common in many bond funds, savings products, and some cash-like investments.
- *Dividends*: some dividends qualify for lower tax rates if certain conditions are met, while others are taxed at ordinary income rates.
- *Realized capital gains*: when you sell an investment for a profit, the gain is taxable. Hold longer than a year and

it may qualify for long-term rates; sell sooner and it's typically taxed at ordinary rates.

- *Distributions you didn't ask for*: some funds distribute capital gains after trading inside the fund. You can owe taxes even without selling shares.

Notice the pattern: taxes show up when money becomes *visible* to the tax system. Visibility happens through payouts and sales. That's why long-term, low-turnover investing can be tax-efficient: it keeps more of the return compounding inside the portfolio instead of constantly "realizing" it.

The Behavioral Trap: Taxes Can Make You Do Something Stupid

Taxes don't just take money. They can distort behavior.

One common trap is avoiding a perfectly good career move or side gig because you heard you'll "lose it all to taxes." That's bucket myth thinking. Another trap is refusing to sell a bad investment because you don't want to "pay taxes." Paying taxes usually means you made a profit. The goal is not to avoid taxes at all costs; the goal is to maximize what you keep *after* taxes while making good decisions with the money.

The worst version of this is when people hold onto a concentrated, risky position because selling would trigger taxes. So they keep the risk. They keep the anxiety. They keep checking the price. They keep telling themselves they're being "tax smart," when what they're actually doing is letting the tax tail wag the investment dog.

A clean rule: don't take extra risk just to postpone a tax bill. Postponing taxes is great when it's part of a coherent plan. It's terrible when it becomes an excuse to avoid making a necessary portfolio decision.

Marginal Rate Thinking: When It's Actually Helpful

Marginal rate thinking is useful when you're making decisions that directly change taxable income. A few real-world moments where marginal rate clarity saves you money and stress:

- *Evaluating pre-tax retirement contributions*: if contributing to a pre-tax account reduces your taxable income, the value of that deduction is tied to your marginal rate.
- *Choosing between a deductible vs. non-deductible expense*: if an expense is deductible, the tax savings depends on your marginal rate.
- *Planning large income events*: bonuses, stock compensation, or a big freelance year can push more dollars into higher buckets, changing the rate on that top slice.

This is the grown-up way to talk about "tax brackets." Not fear. Not superstition. Just: "What's the rate on the next dollar, and how does that change my decision?"

Effective Rate Thinking: The Number That Keeps You Sane

Effective rate thinking is useful for a different job: keeping your overall picture grounded.

If your marginal rate is the cost of the next mile on a toll road, your effective rate is the average cost of the whole trip. It's how you avoid telling yourself dramatic stories like, "The government takes a third of everything I earn," when your overall effective rate might be much lower.

Effective rate thinking also makes it easier to plan. If you know your rough effective rate, you can estimate take-home pay, set realistic savings targets, and avoid the emotional whiplash of "I made $X, where did it go?" (Spoiler: some of it went to taxes, and some of it went to the little invisible taxes inside your investments.)

A Real-World Mini Case Study: The Same Index Fund, Two Different Outcomes

Consider two coworkers, Elena and Sam. Both invest in broadly diversified index funds for the long haul. Neither trades. Neither chases meme stocks. Their investing behavior is almost boring—which is exactly what you want.

The difference: Elena uses a tax-advantaged retirement account for most of her investing. Sam invests mostly in a taxable brokerage account because it felt "more flexible," and retirement accounts felt "locked up."

Over time, both portfolios grow. But Sam keeps paying small annual taxes on dividends and occasional capital gain distributions. Elena's retirement account keeps more of that money compounding inside the account. The gap between them doesn't explode overnight. It widens slowly, then more quickly, like two runners who started side-by-side but one is carrying a backpack.

Sam isn't wrong to value flexibility. Flexibility is real. But the tax cost of flexibility is also real. The point isn't that one account type is morally superior. The point is that *where* you invest can be as important as *what* you invest in, because taxes are part of the return equation.

The most painful version of this story is when Sam is also taking more investment risk than necessary, thinking the bigger swings will "make up for the taxes." Sometimes they do. Sometimes they don't. Either way, it's a stressful way to live.

The Tax-Smart Mindset: Design for After-Tax, Not Before-Tax

The tax-smart investor doesn't obsess over avoiding taxes. They obsess over controlling tax *timing* and reducing tax *leakage*.

Timing is powerful because compounding is powerful. If you can legally defer taxes, you potentially keep more money invested longer. Leakage matters because recurring drags compound in the wrong direction.

This mindset changes the questions you ask:

- Not "How do I pay zero taxes?" but "How do I avoid paying the same type of tax every year unnecessarily?"
- Not "Should I chase higher returns?" but "Should I place the return-generating assets in the accounts where the IRS takes the smallest bite along the way?"

It also changes how you judge investing strategies. A strategy that looks great pre-tax but creates a steady stream of taxable events can end up underwhelming after-tax. A boring, low-turnover approach can look plain on the surface and win underneath

Actionable Takeaway: A Two-Number Check That Improves Every Money Decision

Do this once a year (or anytime your income changes materially):

1. Write down your *marginal tax rate* for federal income taxes: the bracket your next dollar falls into. This is your "decision rate."
2. Estimate your *effective tax rate*: total federal income tax divided by taxable income (use last year's return as a rough guide). This is your "reality rate."

Then use them the right way: use the marginal rate to evaluate moves that change taxable income (like pre-tax contributions), and use the effective rate to keep your overall plan calm, realistic, and free of bracket myths. Once the buckets make sense, the tax system stops feeling like a monster in the dark and starts feeling like a set of rules you can plan around.

5.3 The Alphabet Soup of Wealth: 401(k)s and IRAs

The first time most people open their benefits portal, it feels like being handed a menu written in code: 401(k), Roth 401(k), Traditional IRA, Roth IRA, HSA, match, vesting, rollover. It's easy to respond the way humans always respond to confusion—by doing nothing.

The tragedy is that in the world of building wealth, "doing nothing" often means skipping the closest thing to free money most adults will ever be offered: an employer match. A match isn't a clever strategy or a market prediction. It's a contractual deal: you put in dollars, your employer adds extra dollars, and those extra dollars immediately start compounding right alongside yours. No hot stock on earth reliably competes with a guaranteed 50% or 100% return before your money even touches the market.

The goal here isn't to turn you into a tax attorney. It's to give you a simple operating system for where your next dollar should go—and why.

The Three-Layer Cake: Investment, Account, and Tax Treatment

When you invest, you're making three decisions at once, even if it doesn't feel like it.

First: *what you buy* (a diversified stock fund, a bond fund, a target-date fund). Second: *where you buy it* (a 401(k), an IRA, a taxable brokerage). Third: *how it's taxed* (taxed now, taxed later, or potentially never taxed again on qualified withdrawals).

Most beginner advice obsesses over the first layer: *"Which fund is best?"* But the second and third layers often matter more in real life because they control two forces that quietly decide outcomes: *taxes and behavior*. The right account makes it easier to invest automatically, and it can reduce how much of your returns get siphoned away over time.

So the alphabet soup starts to taste better if you translate it into plain English:

- A *401(k)* is a workplace retirement account. You usually fund it through payroll (automatic, painless), and you may get a match.
- An *IRA* (Individual Retirement Arrangement) is a personal retirement account you open yourself, independent of your job.
- *Traditional* usually means you get a tax break now and pay taxes later.
- *Roth* usually means you pay taxes now and get tax-free qualified withdrawals later.

Those are the levers. Everything else is a variation.

Employer Match: The Only Investment Return You Can Schedule on Your Calendar

A typical match might look like this: "50% match on the first 6% of pay you contribute." If you earn $60,000 and contribute 6% ($3,600), your employer adds 50% of that ($1,800). That's not a maybe. That's not a forecast. That's not "if the market cooperates." It's a benefit you unlock by participating.

Skipping the match is like walking past a table where someone is handing out $20 bills and saying, "No thanks, I'm waiting for a better opportunity."

There are two practical catches to understand:

Vesting. Some employers require you to stay for a certain period before the match money is fully yours. Your own contributions are always yours; the employer's portion might be earned over time. This isn't a reason to avoid the match; it's a reason to read the vesting schedule so you know the rules of the deal.

The match formula. "Dollar-for-dollar up to 3%" is very different from "50 cents on the dollar up to 6%," even though both max out at a 3% of pay match. The point is to contribute enough to get the *full* match, not just "something."

The Order of Operations: Where Your Next Dollar Goes

When decisions pile up, the brain starts bargaining: "I'll invest after I finish paying off everything," or "I'll wait until I know more." A better approach is a short, repeatable sequence that handles the biggest wins first.

The waterfall is intentionally simple. It's designed for real humans with rent, bills, and limited attention—not for people who enjoy optimizing spreadsheets as a hobby.

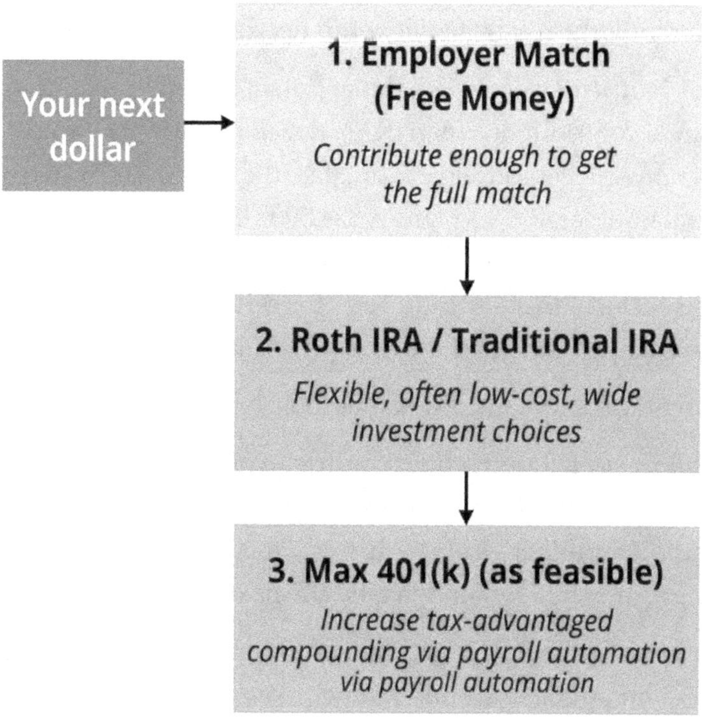

That flowchart is not trying to win an academic debate. It's trying to win your Tuesday night when you're tired and tempted to postpone everything. When you always know what "next" means, progress stops being emotional.

The 401(k): Your Payroll-Powered Wealth Machine

A 401(k) has two superpowers that don't show up in glossy brochures.

Automatic investing. Money leaves your paycheck before it has a chance to become "spendable." That's not a minor convenience; it's behavioral engineering. You don't need willpower every month. You need one setup decision.

High contribution limits. Workplace plans generally allow larger annual contributions than IRAs, which matters once your savings muscle gets stronger. For 2026, the employee elective deferral limit for 401(k) plans is $24,500 (irs.gov). If you're 50 or older, catch-up contributions can raise that ceiling (and there are special higher catch-up rules for certain ages under recent law changes) (irs.gov).

Two practical downsides matter too:

Limited menu. Many plans restrict you to a list of funds. Sometimes the list is excellent. Sometimes it's a museum of mediocre options with high fees. You can still win with a limited menu if you understand the next downside.

Fees. Fees are the silent tax you pay to fund providers. An extra 0.75% in annual fees doesn't feel like much in a single year, but over decades it can siphon off a painful chunk of growth. The most important habit is to find the plan's fund expense ratios and prefer low-cost, broadly diversified options when possible.

If your plan offers a solid low-cost index fund lineup, it's a green light. If it doesn't, you still often contribute enough to get the match (because the match usually overwhelms bad fees), then you become more selective about additional contributions.

IRAs: Your Personal Retirement Account (With More Freedom)

An IRA is what you use when you want more control than your workplace plan offers, or when you want a second tax-advantaged "container" for retirement savings.

For 2026, the annual IRA contribution limit is $7,500 (with additional catch-up allowed for many people age 50+) (irs.gov). That limit is smaller than a 401(k)'s, but the IRA often gives you:

- more investment choices,
- potentially lower-cost fund options,
- and more flexibility in choosing a provider you actually like.

One key nuance: a *Traditional IRA* contribution may or may not be deductible depending on your income and whether you (or your spouse) are covered by a workplace retirement plan. If it's not deductible, it can still be useful in certain scenarios, but it adds paperwork and complexity because you need to track after-tax basis correctly.

That's why many people default to a Roth IRA when eligible: it's simpler, and the tax-free qualified withdrawal feature is emotionally satisfying in a way humans tend to stick with.

Traditional vs. Roth: Pay Taxes Now or Pay Taxes Later

The Roth vs. Traditional question sounds philosophical, but it's basically a timing bet.

Traditional: you typically get a tax break now (lower taxable income today), and you pay taxes when you withdraw later.

Roth: you pay taxes now (no deduction for contributions), and qualified withdrawals later are tax-free.

The tricky part is that nobody knows their future tax rate with certainty. Your income could rise. Tax laws could change. Your retirement lifestyle could cost more than you think. Or you could retire earlier, live on less, and be in a lower bracket later.

So the best decision method is not "always Roth" or "always Traditional." It's choosing based on what's most likely true about *you*.

Question	*Traditional tends to shine when...*	*Roth tends to shine when...*
Taxes today vs. later	Your current marginal rate feels high and you expect lower taxable income later	Your current marginal rate feels low and you expect higher taxable income later
Cash-flow flexibility	You want the tax break now to make saving easier (more money stays in your paycheck)	You can afford to pay the taxes now and want future tax-free withdrawals
Uncertainty comfort	You prefer a known benefit today (deduction)	You prefer future certainty (qualified withdrawals not taxed)

A surprisingly good "middle path" is to diversify your tax treatment: hold some Traditional (pre-tax) and some Roth (after-tax). That way, you're not making one giant, irreversible bet on what future tax rates will look like.

A Real-World Case Study: Free Money, Then a Tax Choice

Consider Dani, age 29, earning $72,000. Dani's employer offers: "100% match on the first 4% you contribute." Dani can invest $400 per month.

If Dani contributes 4% of pay (about $240/month), the employer adds another $240/month. That's $480/month going into retirement investing, even though Dani only "feels" $240 leaving the paycheck. If Dani ignores the 401(k) entirely and instead invests $400/month in a brokerage account, Dani is choosing to pay full price for dollars that could have been subsidized.

So Dani sets 401(k) contributions to capture the full match first.

Now Dani still has $160/month of investing capacity left ($400 - $240). Where does it go?

Dani's 401(k) fund options are fine but not amazing, with a couple of index funds and several high-fee actively managed funds. Dani opens a Roth IRA at a low-cost provider and directs the extra $160/month there, buying a simple diversified fund.

Notice what happened. Dani didn't need a heroic stock pick. Dani didn't need to predict the market. The win came from:

- taking the match,
- choosing an account that fits,
- and using automation to make it repeatable.

That is what "practical wealth-building" looks like: not loud, not complicated, and shockingly effective.

What About Roth 401(k)s?

Many plans now offer a Roth 401(k) option: you contribute after-tax dollars through payroll, and qualified withdrawals later can be tax-free. The appeal is obvious: you get the automation and high contribution limits of a 401(k) with Roth-style tax treatment.

If you're early in your career (when your tax rate is often lower) and you expect your income to rise, Roth contributions can be especially attractive. If you're in peak earning years and the marginal tax rate bite feels heavy, pre-tax contributions can be a relief.

A useful way to decide is to ask a blunt question: *Is the tax deduction today the difference between contributing and not contributing?* If yes, Traditional may be the better behavioral choice because it helps you save more consistently. The best tax strategy on paper is worthless if it makes you quit.

The One Rule That Keeps You From Getting Cute: Capture the Match Before You Optimize

People love to optimize. It's comforting. It feels like control. But if you try to optimize before you take the match, you're polishing a car you haven't purchased.

There's a reason the "employer match" step sits right near the top of the waterfall, highlighted like a neon sign. It's a deal with an immediate payoff. Once you've captured it, then you can start caring about expense ratios, Roth vs. Traditional splits, and the finer points of IRA deductibility.

You'll still make imperfect choices sometimes. Everyone does. The win is setting up a system where your imperfections aren't fatal.

Actionable Takeaway: A 20-Minute Setup That Can Pay You for Decades

Do this with your next paycheck in mind:

1. Find your employer match formula (HR portal or benefits packet) and set your 401(k) contribution to capture the *full* match.
2. Choose *one* default investment inside the account (a low-cost diversified option, often an index fund or a target-date fund) so contributions don't sit in cash.
3. If you can invest more after the match, decide whether your next dollars go to a Roth IRA or additional 401(k) contributions based on one question: *Would you rather*

take the tax break today (Traditional) or lock in tax-free qualified withdrawals later (Roth)? If you can't decide, split it.

When the match is captured and contributions run automatically, the alphabet soup stops being intimidating and starts being what it really is: a set of containers designed to help your money grow up without getting eaten alive by taxes and inertia.

5.4 The Freedom Fund: Investing Beyond $59\frac{1}{2}$

Retirement accounts are like nice dinnerware you only use on holidays: powerful, protected, and a little bit ceremonial. The problem is that life doesn't wait politely until age $59\frac{1}{2}$. Houses get bought at 33. Burnout happens at 41. A parent gets sick at 52. A dream business idea shows up at the worst possible time—which is also the best possible time.

That's why a taxable brokerage account deserves a more flattering name. Call it what it really is: your *Freedom Fund*. It's the place you invest for goals that don't fit inside retirement's lockbox. It's the bridge between "I'm building wealth" and "I can actually use it when I need it."

The catch is that this bridge has tolls. In a retirement account, the tax system mostly waits patiently while your investments compound. In a taxable brokerage account, the IRS is more like a roommate who wants their share of the groceries *as you buy them*. Not always. Not on every dollar. But often enough that ignoring taxes can turn a promising plan into a slow grind.

The good news: you don't need to become a tax wizard. You need a few clean principles and a couple of guardrails that keep you from stepping on the biggest landmines.

Why $59\frac{1}{2}$ Matters (And What It Really Means)

Age $59\frac{1}{2}$ is the age when one of retirement's nastiest "do not touch" signs becomes less scary: the 10% additional tax that often applies to taxable distributions from IRAs taken before that age. After $59\frac{1}{2}$, that additional tax generally no longer applies. (You may still owe regular income taxes on pre-tax withdrawals; the point is the *extra* penalty usually disappears.)

A lot of people take this to mean: "So I can't access retirement money until $59\frac{1}{2}$." That's not fully true. There are exceptions and specialized strategies. But building a plan that depends on exceptions is like building a parachute that only opens if you pull the cord in a very specific way.

A taxable brokerage account is simpler: there's no age gate. You can sell investments and withdraw cash whenever you want. Your costs show up as taxes, not as penalties. And that difference changes the psychology of wealth. It's hard to feel free if your money is trapped behind rules you don't understand.

The Freedom Fund's Job: Flexibility Without Self-Sabotage

A taxable account solves one problem and creates another.

It solves *access*. If you want to retire at 50, take a year off at 38, buy a home at 30, or build a "maybe I quit" buffer, you need money that can be used without asking permission from the tax code.

It creates *temptation*. The same account you can tap for a down payment is also the account you can tap for a vacation that felt "deserved." The Freedom Fund works best when it has a clear purpose and a set of rules you actually follow.

So the real strategy is not "open a brokerage account." The strategy is: *build a taxable portfolio that is tax-efficient, low-drama, and aligned with a mid-term timeline.*

The Tax Reality of Taxable Investing: Your Return Has Tiny Holes in It

In a taxable brokerage account, taxes can show up in a few common ways:

- *Dividends*: some are taxed at preferential rates if they qualify; others are taxed like ordinary income.
- *Interest*: often taxed like ordinary income, which can make some bond-heavy strategies less tax-friendly in taxable accounts.
- *Realized capital gains*: when you sell for a profit, you may owe taxes. Hold investments longer than a year and the gain is generally treated more favorably than a quick flip.
- *Fund distributions*: some funds distribute capital gains due to internal trading, creating taxes even if you didn't sell.

The emotional trap is that these taxes don't always feel like "you paid a bill." Sometimes they show up as a smaller refund.

Sometimes they show up as a bigger tax payment. Sometimes they show up as a 1099 form you didn't expect. But the effect is the same: less money stays invested to compound.

So the Freedom Fund's core idea is simple: *minimize unnecessary taxable events while staying invested in a sensible allocation.*

ETFs vs. Mutual Funds: The Practical Difference in a Taxable Account

Many investors end up using ETFs (exchange-traded funds) in taxable accounts, not because ETFs are magical, but because they often have a structural advantage: they can be more tax-efficient in how they handle internal trading and capital gains distributions.

Mutual funds can be excellent, especially low-cost index mutual funds designed for tax efficiency. But some mutual funds distribute capital gains more frequently, which can create taxable income even if you never touched the account.

The point isn't to start a fund-format holy war. The point is to choose vehicles that let you keep more of your growth compounding quietly. If your Freedom Fund is constantly throwing off taxable distributions, it's like trying to fill a bathtub with the drain half open.

A clean heuristic: in taxable accounts, favor low-cost, broad-market index funds and ETFs with low turnover. They tend to be both diversified and tax-aware by design.

The "Tax-Efficient Ingredients" List (And Why It Works)

If retirement accounts are the place you can hold almost anything without annual tax consequences, taxable accounts reward a more curated menu.

Tax-efficient taxable ingredients often include:

- Broad stock index funds or ETFs (U.S. total market, international total market)
- Municipal bonds for certain high-tax-bracket investors (context-dependent; not universally best)
- Treasury-focused bond strategies in some cases (still taxed federally, often exempt from state tax—details vary)

Tax-inefficient ingredients often include:

- High-turnover actively managed funds that realize gains frequently
- Bond funds that throw off lots of taxable interest (especially when your bracket makes that interest expensive)
- REIT-heavy funds that generate distributions that may not qualify for preferential rates

This is not about shaming any investment category. It's about *location*. Some ingredients belong in certain containers. You're designing a kitchen, not a single meal.

A Concrete Case Study: "Retire at 50" Without Playing Tax Roulette

Meet Priya, 36, who wants the option to stop full-time work at 50. Priya is already contributing enough to capture the employer match and is steadily building retirement savings. But Priya does the math and realizes something uncomfortable: even if retirement accounts grow beautifully, they're not the money Priya wants to live on from ages 50 to $59\frac{1}{2}$. Priya needs a bridge.

So Priya creates a Freedom Fund with a clear role: "This account will fund years 50–$59\frac{1}{2}$ if I choose early retirement, or it will become a flexible cushion for whatever life throws at me."

Priya chooses a simple, tax-aware approach:

- Uses low-cost broad-market equity ETFs as the core holdings.
- Keeps the bond exposure modest in taxable to reduce ordinary-income interest taxation, while holding more bond exposure inside tax-advantaged accounts where it doesn't create annual tax drag.
- Automates a monthly transfer into the brokerage account, treating it like another bill.

Then Priya adds one behavioral rule: "I don't sell from the Freedom Fund for lifestyle upgrades. It's for the bridge or true emergencies."

This rule is the part most people skip. But it's what makes the account a *Freedom* Fund instead of a *Leak* Fund.

Capital Gains: The Taxable Account's Speed Limit

Capital gains taxes are often where taxable investing becomes emotionally charged, because selling can feel like "getting punished" for winning.

A few simple truths keep you grounded:

You only pay capital gains tax when you sell at a profit. If you never sell, the gain isn't realized. (That doesn't mean you should never sell; it means you should be intentional about when you sell.)

Holding period matters. If you sell an investment you've held for more than a year, it generally qualifies for long-term capital gains treatment, which is typically more favorable than short-term.

You control realization. In a taxable account, you have meaningful control over when you sell and how much you realize each year. That control is a planning advantage.

So the Freedom Fund isn't just "invest outside retirement." It's "invest with an awareness of when selling creates a tax event."

A practical habit: if you're planning a major withdrawal (down payment, sabbatical), avoid selling everything on impulse. Map the timeline. Decide what you'll sell and when, ideally giving yourself the option to prioritize long-term gains and manage the tax impact.

Tax-Loss Harvesting: Turning Losses Into Future Fuel (Without Breaking the Rules)

Tax-loss harvesting is the taxable account's most misunderstood tool. It sounds like a gimmick. It's not. It's simply using a loss—which feels bad anyway—to reduce taxes.

In basic terms:

- If you sell an investment for less than you paid, you realize a capital loss.
- Capital losses can offset capital gains.
- If losses exceed gains, you may be able to deduct up to $3,000 against ordinary income in a year (subject to rules), and carry forward the remaining losses to future years.

The crucial rule is the *wash sale*. If you sell a security at a loss and buy the same or "substantially identical" security within 30 days before or after the sale, the loss is generally disallowed for current tax purposes and is instead added to the basis of the replacement shares (the loss is typically deferred, not erased). This is where people accidentally step on rakes.

A realistic, non-technical way to use this: if you harvest a loss in one ETF, you can buy a similar—but not substantially identical—ETF to maintain market exposure. The goal is to keep your long-term allocation intact while using the tax code's rules to reduce the drag.

You do not need to do this monthly. You do not need to chase losses. It's a tool, not a personality.

Dividends: The "Nice Surprise" That Can Be Tax-Expensive

Dividends feel friendly. You get paid for holding an investment. It feels like progress.

But in a taxable account, dividends can create taxes even if you reinvest them immediately. And not all dividends are taxed the same way. Some qualify for preferential tax rates if certain requirements are met; others are taxed like ordinary income.

This doesn't mean you should avoid dividends entirely. It means you should stop treating dividend yield like a free lunch. A high-dividend strategy in a taxable account can be less tax-efficient than a total-return approach that lets growth compound quietly and gives you more control over when to realize gains.

The Freedom Fund loves control. It hates surprises.

A Simple Freedom Fund Blueprint (That Respects Your Dinner Plate)

Asset allocation still matters. The Freedom Fund is not a permission slip to invest like a gambler. The key difference is that the Freedom Fund's timeline is often mid-term, so it needs a balance between growth and stability that matches when you plan to use it.

A practical way to design it:

- *If the goal is 10+ years away*: you can often keep a meaningful stock allocation for growth, with some stabilizers if that helps you stay invested.

- *If the goal is 3–7 years away*: consider dialing down volatility. You can still invest, but you don't want your down payment depending on a market recovery schedule.
- *If the goal is under 3 years*: the Freedom Fund starts to look less like "investing" and more like "capital preservation." Cash and short-duration, high-quality options become more relevant.

This is the glide-path idea in street clothes: as the deadline gets closer, you reduce the chance that bad timing ruins a good plan.

The Most Dangerous Freedom Fund Mistake: Treating It Like a Checking Account

The taxable brokerage account's flexibility is its superpower—and its biggest risk.

The temptation is to treat it as "extra money," because it's visible, accessible, and not wrapped in retirement language. That's how people end up selling investments for vacations, weddings they couldn't afford, or lifestyle upgrades that feel justified because "it's not retirement money."

If you want the Freedom Fund to deliver actual freedom, it needs a boundary. A boundary isn't restrictive; it's clarifying. It tells Future You, "I protected this money from Present Me."

A good boundary is specific:

> *I only withdraw from the Freedom Fund for (1) the named goal, (2) a true emergency after my cash emer-*

gency fund is exhausted, or (3) a planned life transition I've given at least 60 days of thought.

Sixty days sounds arbitrary, but that's the point. It creates a speed bump. Most impulse spending can't survive a speed bump.

Actionable Takeaway: Build a Freedom Fund That Actually Buys Freedom

Do three things this week:

1. Open a taxable brokerage account (if you don't already have one) and nickname it *Freedom Fund* with a one-sentence purpose: "This is for ______________ by __________."
2. Choose a tax-efficient default investment for automatic contributions (typically a low-cost, broad-market index fund or ETF) and set an automatic monthly transfer.
3. Write one withdrawal rule you'll follow and keep it where you can see it: "I don't sell from this account for lifestyle upgrades."

That's how a taxable brokerage account stops being "the place where my money sits" and becomes what it's meant to be: a flexible, tax-aware reservoir of options, built patiently enough that you can make big life decisions without asking your employer, your future self, or the tax code for permission.

5.5 The Art of Rebalancing

There's a moment every investor runs into, usually after their first real bull market, when they open their account and feel like a genius. The numbers are up. One fund is up *a lot*. Your brain starts telling a story: "Clearly I'm the kind of person who picks winners," or the more subtle version, "Maybe I should just let the winners run forever."

Then reality does its favorite thing: it changes the weather.

Rebalancing is the antidote to the two most expensive emotions in investing: *pride* and *panic*. It's the maintenance work that keeps your portfolio's risk from quietly mutating behind your back. And it has a weird job description: you sell a little of what has been good to you and buy a little of what has disappointed you. It feels backward. It feels like breaking a streak. It feels like telling the market you don't appreciate its generosity.

It's also one of the cleanest ways to make sure your portfolio stays aligned with the dinner plate you chose on purpose, rather than the one the market builds for you by accident.

Portfolio Drift: When Your ``60/40" Becomes a Different Person

When people say, "My portfolio is 60% stocks and 40% bonds," it sounds like a permanent trait, like eye color. In reality it's more like posture. You don't set it once and keep it forever. It slumps.

If stocks have a great year and bonds are flat, your stocks become a larger slice of the pie. You didn't choose more risk; you *drifted* into it. Your portfolio's behavior changes even if you never touch a button.

That drift is sneaky because it feels good at first. A rising market makes your growth assets swell. Your account balance climbs. The thing that's changing—your risk—isn't printed in fireworks.

But risk has a way of introducing itself dramatically later.

A portfolio that drifts from 60/40 to 75/25 can become much more volatile than you originally intended. That might be fine for someone with a long horizon and a strong stomach. It can be disastrous for someone building a Freedom Fund meant to be tapped in a few years, or for someone who chose 60/40 specifically because they wanted the account to feel stable enough to hold through ugly headlines.

Rebalancing is how you correct posture.

Rebalancing Is Not Market Timing (It's Risk Management)

Market timing is when you try to predict short-term moves: "Stocks are going down, so I'm selling now and buying back later." That strategy requires you to be right twice: when to get out and when to get back in. Most people struggle to be right once.

Rebalancing is different. Rebalancing says: "I don't know what the market will do next. I'm simply returning to my chosen mix." It doesn't require forecasts. It requires a rule.

That rule is important because the human brain is spectacularly unreliable at staying calm when money is involved. When markets soar, you feel invincible and take more risk. When markets crash, you feel doomed and take less risk. Left to instinct, most people do the investing equivalent of buying high and selling low, just with better excuses.

Rebalancing builds "buy low, sell high" into a routine. Not as a cute slogan, but as an automatic consequence of returning to target weights.

The Counterintuitive Magic: Selling Winners Without Hating Them

People resist rebalancing because it sounds like punishment for success. "Why would I sell the thing that's working?"

Because your goal is not to own the best-performing asset in a specific year. Your goal is to finance a life.

A portfolio isn't a scoreboard. It's a machine designed to deliver outcomes with tolerable stress. That requires a balance of ingredients. If one ingredient swells to dominate the plate, the meal becomes unbalanced, and so does your risk.

A useful mental trick: don't think of rebalancing as "selling the winner." Think of it as "skimming the foam." When a pot boils over, you reduce the heat and bring it back under control. You're not angry at the pot. You're preventing a mess.

A Concrete Example: The 60/40 That Quietly Turned Into 72/28

Say you start with $100,000:

- $60,000 in a broad stock fund
- $40,000 in a bond fund

Then the market does what it does. Stocks rally 30%. Bonds earn 2%.

Now you have:

- Stocks: $60,000 × 1.30 = $78,000
- Bonds: $40,000 × 1.02 = $40,800

Total: $118,800.

Your new allocation is roughly:

$$\text{Stocks} \approx 65.7\%, \quad \text{Bonds} \approx 34.3\%.$$

That drift isn't wild yet. But imagine the same pattern repeats for several years, or imagine a decade like the late 2010s when stocks outpaced bonds by a wide margin. You can end up with a portfolio that behaves like an aggressive investor's portfolio even if you never meant to be aggressive.

Rebalancing pulls it back:

$$\text{Target stocks} = 0.60 \times 118{,}800 = 71{,}280$$

$$\text{Target bonds} = 0.40 \times 118{,}800 = 47{,}520$$

So you would sell about $6,720 of stocks ($78,000 - $71,280) and buy about $6,720 of bonds.

That move doesn't predict anything. It simply restores the portfolio's original character.

The Real Reason Rebalancing Works: It Keeps You From ``Accidentally" Changing the Plan

Most investors don't wake up and say, "I'd like to become more aggressive with my retirement savings at exactly the moment stocks are expensive." But that's what drift does. It quietly increases your exposure to what has already gone up.

Rebalancing is how you avoid the accidental plan change.

It also protects you from the reverse mistake. When stocks fall, your stock allocation shrinks as a percentage of the portfolio. If you do nothing, you become more conservative after a decline—which is exactly when future expected returns may be more attractive. Rebalancing forces you to buy when the thing is on sale, not when it's trending on social media.

This is why rebalancing feels emotionally unpleasant: it often pushes against the mood of the crowd.

Two Simple Rebalancing Methods (Both Legit)

There are two mainstream ways to rebalance. Pick one. The best method is the one you'll actually do.

Calendar rebalancing

You rebalance on a schedule: every 6 months, every year, or once a year on your birthday. It's simple, predictable, and requires minimal attention.

Calendar rebalancing is especially good for people who want investing to be a low-mental-load habit. You don't check constantly. You check when the calendar tells you to.

Threshold (drift-band) rebalancing

You rebalance when your allocation drifts beyond a set threshold—for example, if stocks drift more than 5 percentage points away from target.

This method can be more responsive in turbulent markets. It also reduces unnecessary trading when markets are calm.

The downside is that it requires you to monitor your allocation, which can tempt some people into fiddling too often. If you know you're a fiddler, choose the calendar method and protect yourself from yourself.

The Cleanest Way to Rebalance: Use Cash Flows First

Selling can be psychologically hard. In taxable accounts, selling can also create capital gains taxes. So whenever possible, rebalance with *new money*.

If your target is 60/40 and stocks have run up, your next contributions can be directed more toward bonds until you're back near target. This can correct drift without selling anything.

Similarly, if you're withdrawing (for example, from a Freedom Fund), you can withdraw from the asset class that is overweight. That's rebalancing in reverse, and it's elegant. You're removing the excess without triggering extra trades.

Cash-flow rebalancing is the quiet, low-friction option. You still might need to sell occasionally, but you can often reduce how much selling is required.

Case Study: Rebalancing During a Scary Year

Meet Luis, 42, who invests in a simple two-fund portfolio: U.S. stocks and high-quality bonds. Luis isn't trying to retire tomorrow, but Luis is also not interested in stomach ulcers.

Luis chooses 70/30 because it feels like enough growth without making every market dip feel like a personal crisis. For a few years, stocks do well. The portfolio drifts to 80/20. Luis notices and thinks, "Maybe I'm just more aggressive than I thought."

Then a bad year hits. Stocks drop sharply. Bonds are steadier. Luis opens the account and sees the portfolio is down, but not destroyed. The new allocation is closer to 72/28 because stocks fell.

Luis has a choice:

- Follow instinct: "Stocks are dangerous" and reduce stock exposure.
- Follow the rule: rebalance back to 70/30 (which may require *buying* stocks while everyone else is scared).

Luis follows the rule. Not dramatically. Not heroically. Just mechanically.

Months later, when markets stabilize and recover, Luis is glad the rule existed. The rule didn't guarantee a profit. It guaranteed something more valuable: Luis didn't change identity mid-storm. The portfolio stayed the portfolio Luis had chosen with a calm brain, not the portfolio Luis would have chosen with an adrenaline brain.

That's what rebalancing is really for.

Rebalancing in Taxable Accounts: Don't Turn Risk Control Into a Tax Disaster

In a retirement account, rebalancing is usually straightforward: you can buy and sell within the account without creating a current-year tax bill. That makes retirement accounts a natural home base for rebalancing moves.

In taxable accounts, selling appreciated investments can realize capital gains. That doesn't mean you can't rebalance. It means you should do it thoughtfully.

A few practical principles:

- *Prefer cash-flow rebalancing* (directing new contributions) when possible.
- *Use tax lots intentionally* if your brokerage supports it (selling shares with higher cost basis can reduce realized gains).
- *Harvest losses when available* to offset gains, while respecting wash-sale rules.
- *Rebalance less often* unless drift is meaningful. The goal is risk control, not hyperactivity.

Risk management that triggers a big avoidable tax bill is like buying a helmet so heavy you fall over. Protective, but not practical.

When Not to Rebalance

Rebalancing is a tool, not a reflex. There are times when "do nothing" is the correct move.

Tiny portfolios. If your portfolio is small and you're contributing regularly, new contributions can do almost all the rebalancing work.

Tiny drift. If you're off target by a percent or two, the benefit of trading is usually not worth the hassle.

A genuinely changed goal. If your timeline changes—you're buying a house sooner, retiring earlier, changing careers—then the right move may be to change the *target allocation*, not to rebalance to an old target that no longer fits your life.

Rebalancing is about staying faithful to your plan. If the plan needs to change for a real reason, change it intentionally. Don't let drift make the decision for you.

A Practical Rebalancing Rule You Can Actually Live With

If you want something simple and sturdy, use this:

> *Once per year, I will check my allocation. If any major asset class is more than 5 percentage points away from target, I will rebalance back toward target, using new contributions first and trades only if necessary.*

That rule is calm. It's not obsessive. It's also strong enough to matter.

You can also tie the check to something memorable: a birthday, a New Year's ritual, tax season, or the month your annual bonus arrives. The key is to make it part of life, not a separate hobby you forget.

Actionable Takeaway: Write Your Rebalancing Policy in One Sentence

Before the market forces you to make decisions while stressed, decide your rule now. Write it somewhere you'll see it:

> *My target allocation is ________/________. I rebalance ________ (annually / when drift exceeds _____%). I use new contributions first and avoid taxable selling unless drift is meaningful.*

The real art of rebalancing is not the math. The math is easy. The art is choosing a rule you can follow when your portfolio is up, when it's down, and when everyone around you is suddenly an expert with very loud opinions.

5.6 Pushing the Button: Execution and Automation

A financial plan that lives in your head is a wish. A plan that lives in your calendar and your payroll system is a machine.

Most people don't fail at investing because they picked the wrong index fund. They fail because their investing requires them to be a calm, responsible person *every single month*. The rent is due, a tire blows out, someone invites you to a wedding in a city with $18 cocktails, and suddenly "I'll invest later" becomes a personality trait.

Execution is where you turn the nice ideas—asset mix, tax awareness, account hierarchy—into something that happens while you're busy being human. Automation is how you make it happen even when you're tired, distracted, or convinced the market is about to do something dramatic.

The goal is not to become a trading expert. The goal is to place a few well-chosen dominoes so your wealth-building keeps moving without you constantly pushing it.

5.7 Pick Your Platforms Like You Pick Your Gym: Close, Simple, and One You'll Actually Use

People love debating which brokerage is "best," the way they debate which gym is "best." Then they never go.

A modern investing setup usually involves one or more of these:

- *Workplace plan* (401(k), 403(b), etc.): the payroll-powered engine, often with a match.
- *IRA provider*: your personal retirement account, often with more fund choices.
- *Taxable brokerage*: the Freedom Fund for goals before $59\frac{1}{2}$.
- *Robo-advisor* (optional): a service that picks a diversified portfolio and often rebalances automatically for a fee.

The best platform is one that supports three behaviors:

1. *automatic contributions* (from payroll or scheduled bank transfers),
2. *low-cost diversified investing* (index funds/ETFs, target-date funds, or similar),

3. *low friction* (good interface, easy customer service, straightforward statements).

If a platform makes automation difficult, it's the wrong platform for the job. You're not shopping for "features." You're shopping for fewer excuses.

5.8 Open the Account: The Tiny Bureaucracy That Changes Your Life

Opening an account is rarely hard. It's just irritating. It asks you for the kind of details you don't keep in your brain on purpose: Social Security number, employer address, bank routing number, beneficiary info.

That irritation is a filter. It's the universe asking, "Do you want this badly enough to be mildly inconvenienced for 15 minutes?"

Here's the clean way to power through:

- Set a 30-minute timer.
- Gather: ID, SSN, bank info, employer info, and a list of beneficiaries.
- Open the account and *immediately* link your bank or payroll contribution.

The last bullet is the difference between an account that exists and an account that works. People open accounts all the time and then never fund them. That's like buying a car and leaving it in the driveway without gas.

5.9 Choosing the First Investment: Boring Is a Feature, Not a Bug

Once the account is open, you face the most emotionally loaded dropdown menu in personal finance: the list of investments.

This is where people freeze. The options look like a wall of abbreviations and numbers. Your brain starts begging you to do research. Research feels responsible. Research is also an excellent hiding place.

A practical default is to choose a single diversified option that matches your asset allocation:

- In many workplace plans: a *target-date fund* can be a perfectly reasonable "set it and keep moving" choice, especially if it's low cost.
- In an IRA or brokerage: a *broad-market index fund or ETF* (or a simple two- or three-fund mix) can serve as a stable core.

The real skill is not picking the perfect fund. The real skill is picking a good-enough fund and then making contributions automatic so time does the heavy lifting.

5.10 Target-Date Funds: The Autopilot You Still Need to Glance At

Target-date funds sound like magic: pick the year you expect to retire, and the fund gradually shifts from stocks toward bonds over time. That gradual shift is called a *glide path*.

The benefit is behavioral: you're less likely to tinker. The fund rebalances and adjusts risk for you.

The caution is that target-date funds are not identical. Their glide paths differ. Their fees differ. Their underlying holdings differ. In some workplace plans, the target-date option may be a collective investment trust (CIT) rather than a mutual fund, which can affect what disclosures you see and how information is presented.

So the right posture is: trust the autopilot, but check the dashboard once in a while. Specifically:

- Look at the expense ratio (lower is generally better, all else equal).
- Confirm the stock/bond mix roughly matches your risk appetite.
- Make sure you understand whether it's designed to land at the retirement year and stop, or to keep getting more conservative afterward.

If this sounds like a lot, remember: you're not analyzing 200 stocks. You're checking three dials.

5.11 The Trade Itself: Market Order vs. Limit Order (Without the Drama)

Most long-term investors don't need to become traders, but you do need to understand the two buttons that can change outcomes when you buy or sell:

Market order: "Buy/sell this now at the best available price." You get speed and convenience. You give up control over the exact price.

Limit order: "Buy/sell only at this price or better." You get price control. You risk not getting filled if the market moves away.

For liquid, broad-market ETFs during normal market hours, a market order is often fine for small, long-term purchases. The risk of price whiplash is generally low, though not zero. For less liquid investments, larger trades, or moments of high volatility, limit orders can prevent unpleasant surprises.

A clean rule for real life:

> *If I'm investing for the long term in a highly liquid fund and the trade is not huge relative to my portfolio, a market order during regular hours is usually okay. If I care about the exact price, I use a limit order.*

Also: avoid placing market orders right at the open or close if you can help it. Those moments can be more volatile. This isn't superstition; it's about letting the market settle.

5.12 Dollar-Cost Averaging: The Most Misunderstood Superpower

Dollar-cost averaging is simply investing a fixed amount on a schedule—say $300 every month—regardless of what the market did recently. It's what most workers already do in a 401(k): every paycheck, money buys whatever the plan is set to buy.

People sometimes mistake dollar-cost averaging for a way to "beat the market." It's not. It's a way to beat *yourself*.

It helps because:

- When prices are high, your fixed contribution buys fewer shares.
- When prices are low, it buys more shares.

The real benefit is that it prevents you from waiting endlessly for "the perfect time," which tends to arrive right after the market has already recovered.

If you have a lump sum and you're deciding whether to invest all at once or spread it out, that's a separate decision with trade-offs. But for most people building wealth from paychecks, consistent automatic investing is the simplest path that still works.

5.13 Automation: The Difference Between ``I Intend To" and ``I Do"

Automation isn't a convenience; it's a strategy.

The best investing plan in the world loses to a mediocre plan that happens every month.

Set up automation in three places if possible:

5.14 1) Payroll contributions (workplace plan)

If your employer offers retirement contributions through payroll, use it. It is the least friction you will ever experience in investing. The money never becomes "available" to spend, which means you never need to fight yourself for it.

5.15 2) Automatic bank transfers (IRA and taxable brokerage)

Pick a date that aligns with your cash flow, usually 1–3 days after payday. Schedule a transfer to your IRA or Freedom Fund.

Your goal is not to time the market; your goal is to time your *behavior*. You want the transfer to happen before money gets assigned to random spending.

5.16 3) Automatic investing inside the account

Some brokerages let you automatically invest contributions into specific funds/ETFs. If you can automate the "buy" step, do it. If you can't, set a recurring calendar reminder. It's still a system; it's just semi-automatic.

5.17 Case Study: From ``I'll Start Soon" to a Working System

Meet Tasha, 27, who has been "meaning to invest" for two years. Tasha reads articles, listens to podcasts, and has a notes app full of fund tickers. Tasha is not lazy. Tasha is overwhelmed.

Tasha's turning point isn't a brilliant insight about markets. It's a Tuesday.

At lunch, Tasha opens the benefits portal and finds the match: the employer will match 50% up to 6% of pay. Tasha sets the contribution to 6%. That's the first domino.

That night, Tasha opens a Roth IRA and sets a $200 monthly transfer for the day after payday. Tasha chooses a single diversified fund. No sprawling menu. No endless comparisons.

The final domino is embarrassingly small: Tasha names the brokerage account "Future Me" and writes one sentence in the account notes: *"This is not for vacations."*

Six months later, Tasha hasn't become a different person. The market has been weird. Work has been busy. Tasha has not become more disciplined in any heroic way. But the balances are growing because the system runs even when motivation doesn't show up.

That's what "execution" looks like in real life: a few decisions that remove the need for daily virtue.

5.18 Robo-Advisors: When Paying for Automation Makes Sense

A robo-advisor can be worth it if it solves the two problems that trip you up:

- You struggle to pick a portfolio and stick with it.
- You struggle to rebalance or manage taxes in a taxable account.

Robo-advisors typically charge an advisory fee on top of the underlying fund expenses. That fee buys you convenience and discipline. If the alternative is doing nothing, or constantly changing strategies, the fee can be a bargain.

If you're comfortable holding a simple index portfolio and rebalancing occasionally, you may not need a robo-advisor. The point isn't to avoid fees as a moral mission. The point is to pay for things that genuinely improve your behavior and outcomes.

5.19 The Safety Checklist: The Boring Steps That Prevent Expensive Mistakes

Execution also includes a few "adulting" moves that feel unglamorous until they save you:

- *Set beneficiaries* on retirement accounts. This is one of the simplest estate-planning actions you can take.
- *Turn on two-factor authentication* for your brokerage and email accounts. Investing is digital; so is theft.

- *Know where fees are listed* in your workplace plan and what your fund expense ratios are. You don't need to obsess, but you do need awareness.
- *Keep an emergency fund separate* so you don't raid investments for predictable life events.

None of these steps make you richer overnight. They make you harder to derail.

5.20 Actionable Takeaway: Build a One-Page Investing Routine

Write this down and make it your default operating system:

1. *Today:* Set 401(k) contributions to capture the full match (if available) and pick a default investment.
2. *This week:* Open an IRA or taxable brokerage (as appropriate) and schedule an automatic transfer for the day after payday.
3. *Monthly:* If buys aren't automated, place them on the same date each month. Same funds. Same plan. No debating.
4. *Yearly:* Review allocations and fees, update beneficiaries if needed, and rebalance if you've drifted meaningfully.

Once those routines are in place, wealth-building stops being a motivational project and becomes something closer to brushing your teeth: small, automatic actions that quietly protect Future You from the chaos of regular life.

Chapter 6

Mastering Psychology and Life Decisions

If spreadsheets were enough, we would all be billionaires with six-pack abs. The truth is, personal finance is 20% math and 80% behavior. This chapter explores why our brains are hardwired to sabotage our bank accounts and how to rewire them. We will move beyond the latte-shaming to tackle the massive decisions that actually move the needle: choosing whether to rent or buy, understanding the true cost of a mortgage, and recognizing that your career is your single greatest asset.

6.1 Rewiring Your Money Mindset

At 2:17 a.m., someone in sweatpants is doing math on a glowing screen they don't fully trust. The market is down. Their index fund looks like it has a leak. Every red number whispers the same ancient message: *Get out before it gets worse.* They aren't thinking about retirement anymore. They're thinking about a lion in the grass.

The uncomfortable truth is that your brain is not designed to build wealth. It's designed to keep you alive long enough to reproduce. For most of human history, "good decisions" meant eating when food appeared, fitting in with your group, and avoiding danger fast. Wealth-building asks you to do the opposite: delay gratification, tolerate uncertainty, ignore status contests, and stay calm while your account balance bounces around like a heartbeat monitor.

Money makes this harder because it's not just numbers. It's safety, identity, freedom, and shame jammed into a spreadsheet. When people say "I'm bad with money," they usually mean "My nervous system takes over." If you want to change your finances, you don't start by downloading a budgeting app. You start by learning the handful of brain glitches that reliably hijack smart people.

Your brain runs two programs

You have a fast system and a slow system. The fast one is automatic: it reacts, guesses, protects, compares. It's the part that flinches when a car swerves, refreshes your portfolio during a market dip, and adds "just one more thing" to the cart because it's on sale. The slow one is deliberate: it plans, calculates, and can sit with discomfort long enough to make a better choice.

Here's the catch: the fast system speaks first and loudest. The slow system is quieter and gets tired. After a long day, the fast system effectively has the microphone. That's why "I'll decide tonight" is a dangerous sentence in personal finance. Nighttime decisions are where overpriced cars, bad loans, and panicked trades are born.

The goal isn't to delete the fast system. It's kept humans alive for a very long time. The goal is to recognize when it's driving and install speed bumps that force the slow system to show up before you do something expensive.

Loss aversion: why losing $100 hurts like finding $200 feels good

If you've ever sold an investment at the worst possible time and then watched it recover, you've met *loss aversion*. Psychologists have found that losses feel roughly twice as painful as equivalent gains feel pleasurable. Your brain treats a loss like a threat. A threat demands action.

This makes perfect evolutionary sense. In the wild, a "loss" could mean injury, social rejection, or starvation. Overreacting to losses kept your ancestors from becoming lunch. Underreacting didn't.

Modern money flips the logic. Investing requires you to accept frequent, small "losses" on paper (short-term drops) in exchange for long-term gains. Your account balance becomes a threat detector, and the market happily provides false alarms.

Loss aversion also explains why people cling to broken financial ideas long after the math says they're wrong. A homeowner keeps a house they can't afford because selling would "lock in" the loss. An investor refuses to rebalance because selling a winner "feels like" giving up future upside and selling a loser "admits" failure. A worker stays in a dead-end job because leaving means risking the familiar paycheck for the uncertain one.

A simple reframe helps: a market decline is not a predator; it's a price. The pain you feel is not a signal that you're in danger. It's your brain's ancient warning system reacting to a modern situation it doesn't understand.

A real-world case: the 2020 whiplash and the cost of ``doing something"

Consider what happened to millions of ordinary investors in early 2020. Markets fell fast. News headlines sounded apocalyptic. Many people sold because their brains screamed, *This is different. This is permanent.* Some moved to cash "until things settled down."

Then the rebound came with brutal speed. The investors who sold didn't just miss returns. They missed the recovery days that often drive a huge portion of long-term gains. Many didn't buy back in until prices were higher, meaning they converted a temporary drop into a permanent wealth haircut.

The tragic part is that their decision felt responsible. It felt like "risk management." That's loss aversion wearing a suit. It turns panic into a virtue.

You don't need to memorize market history to learn from this. You only need one rule: when fear is high, your desire to take action spikes, and the action is usually the wrong one. In investing, "doing something" is frequently just paying tuition to your nervous system.

Social comparison: the Joneses aren't rich, they're leveraged

If loss aversion is the fear trigger, social comparison is the status trigger. Humans are social animals. In a tribe, knowing your rank mattered: it influenced access to mates, food, and protection. So your brain constantly asks, *How am I doing relative to them?*

Marketers know this. Entire industries exist to translate insecurity into purchases. A neighbor's new SUV doesn't just look nice; it subtly threatens your identity. A friend's luxury vacation doesn't just look fun; it implies your life is smaller. Social media turns this into a 24/7 highlight reel, and your brain responds as if it's receiving real-time status updates from your tribe.

The cruel trick is that you never see the full balance sheet. You see the kitchen remodel, not the credit card debt. You see the leased car, not the monthly payment. You see the wedding, not the loan. You see the outfits, not the empty retirement account.

Plenty of people with expensive-looking lives are broke in the most modern way possible: not poor, just *fragile*. One job loss away from panic. One medical bill away from a high-interest spiral. The Joneses are often not wealthy; they're financed.

Wealth is quiet because it's frequently invisible. It looks like driving a paid-off car by choice. It looks like saying no to upgrades that don't matter. It looks like having cash reserves and boring index funds, which is not exactly content-friendly.

The "reference point" trap: why you adapt to raises and still feel behind

A sneaky feature of the human brain is that it judges outcomes relative to a reference point, not in absolute terms. That means your satisfaction doesn't come from what you have; it comes from what you *expected*.

Get a $10,000 raise and you might feel rich for a month. Then your brain resets the baseline. Your new paycheck becomes "normal," and the old paycheck becomes unthinkable. This is why lifestyle inflation feels so natural. It's not just greed. It's a reference-point upgrade.

The same thing happens with homes. A slightly bigger place becomes the new normal, and suddenly the old place feels like deprivation even if it was perfectly fine. The baseline shifts. Your brain treats the new level as the minimum acceptable life. That's how people earn more and feel poorer.

The antidote isn't guilt; it's intentionality. You can decide which upgrades become permanent and which remain occasional treats. You can set a reference point on purpose: *My baseline is stability and freedom, not appearances.*

Present bias: why ``future you" keeps getting stuck with the bill

Present bias is the tendency to overweight what happens now and underweight what happens later. Your brain discounts future rewards because, historically, the future was uncertain. Planning a feast next month didn't help if you didn't survive the week.

This shows up in money as procrastination and "small" decisions that snowball. You put off enrolling in the 401(k) because the forms are annoying. You delay switching insurance even though it would save hundreds a year. You keep paying for subscriptions you don't use because canceling is mildly unpleasant. You accept a high-interest rate because shopping feels like work.

Present bias also explains why people can be genuinely intelligent and still run up credit card debt. The purchase gives an immediate hit of pleasure or relief. The payment is delayed and abstract. The pain is future-you's problem. Your brain is basically running a tiny payday loan against your own tomorrow.

To fight present bias, you need to make the future feel real. Not as a vague "retirement," but as a specific person with a life. One of the most effective mindset shifts is to treat future-you like a teammate you respect. You wouldn't steal from a coworker's paycheck. Don't steal from your own.

The "money scripts" you inherited without consenting

Most money beliefs aren't taught explicitly. They're absorbed. You watched how adults argued about bills. You noticed whether money was discussed openly or treated like a shameful secret. You saw whether spending was used as comfort, whether debt was normal, whether talking about income was taboo.

Those early impressions become scripts:

- "Rich people are greedy."
- "Debt is just how life works."
- "If I can't buy it twice, I can't afford it" (useful for some, paralyzing for others).
- "I deserve this" (sometimes true, sometimes a disguised coping mechanism).

Scripts aren't facts. They're stories your brain repeats because it prefers a familiar narrative to uncertainty. If you never challenge them, you will unconsciously build your financial life around them.

A practical way to surface a script is to finish this sentence in writing: *People like me don't...* Invest? Negotiate? Earn a lot? Talk about money? Own nice things? Start businesses? Whatever comes out is the belief running in the background. Once you can see it, you can test it.

Install a ``pause button" for impulse purchases

Impulse spending isn't a character flaw. It's a predictable response to emotion plus frictionless purchasing. When stress is high and buying is easy, your brain reaches for relief.

So the fix isn't motivational speeches. It's engineering.

A pause button is any rule that inserts time between the urge and the purchase, long enough for your slow system to wake up. The point is not to *never* buy fun things. The point is to stop buying things you don't actually want with money you'll miss later.

Here are three pause buttons that work because they're simple and annoying in the right way:

- *The 24-hour rule for anything nonessential over $50.* Put it in a notes app or cart, then walk away. If you still want it tomorrow, you're allowed to decide with a cooler brain.
- *The "cash explanation" rule.* Before buying, write one sentence: "I'm buying this because __________." If the hon-

est answer is "I'm stressed" or "I feel behind," you've caught social comparison in the act

- *Friction by default.* Remove saved cards from websites, disable one-click buying, and keep shopping apps off your home screen. You're not making life harder; you're making your future easier.

This works because your fast system hates friction. The slow system benefits from it.

A calmer relationship with risk: volatility is the price of admission

People often say they want "safe" investing. What they usually mean is "I want to make money without feeling bad." Unfortunately, the market doesn't charge fees in dollars; it charges fees in emotion. Volatility is the price of admission for long-term returns.

If you want the upside of ownership in businesses (stocks), you pay with temporary declines. If you can't tolerate that, you'll either avoid investing (and lose to inflation) or you'll invest and then sabotage yourself at the exact wrong moments.

A helpful mental model is to separate *risk* from *volatility*. Volatility is movement. Risk is the chance of permanent loss or the inability to meet your goals. A diversified long-term portfolio can be volatile without being "risky" in the way your brain imagines at 2:17 a.m.

The mindset shift is to stop interpreting red numbers as a personal failure. A down month isn't an indictment. It's weather.

Make your identity about behaviors, not outcomes

One reason money feels so emotional is that people fuse it with self-worth. A good month means you're smart. A bad month means you're irresponsible. That's a recipe for shame, and shame is gasoline for avoidance.

A more stable identity is behavior-based:

- "I am the kind of person who pays myself first."
- "I am the kind of person who checks fees and rates."
- "I am the kind of person who waits 24 hours before big purchases."
- "I am the kind of person who invests consistently, especially when it feels boring."

Outcomes bounce around. Behaviors compound.

You can be a person with a strong financial identity even if you're starting from zero, even if you've made mistakes, even if your income is modest. Wealth is not a personality trait. It's the accumulated result of a small set of repeated actions done for a long time.

Actionable takeaway: a 10-minute ``mindset audit" you can do today

Set a timer for 10 minutes and write answers to these prompts on paper (not in your head):

1. *My most expensive money emotion is:* stress / boredom / envy / fear / guilt / something else.

2. *When that emotion shows up, my default money move is:* spend / avoid / overwork / gamble / freeze.
3. *The cognitive trap behind it is most likely:* loss aversion (panic) / social comparison (status) / present bias (procrastination).
4. *My pause button will be:* a 24-hour rule, removing saved cards, or a written "cash explanation" sentence before buying.
5. *One sentence I want future-me to be able to say is:* "Thank you for ______________."

Then do one physical action immediately: delete one shopping app, remove one saved card, set one automatic transfer, or write your 24-hour rule on a sticky note and put it where you usually spend. Your brain will keep producing old urges. The difference is that now they have to pass through a gate you designed.

6.2 Building Unbreakable Habits

At 8:43 p.m., your bank app feels like it weighs twenty pounds.

All day you made reasonable choices. You answered messages, handled problems, navigated other people's moods, resisted three snacks you didn't need. Then you sit down to "finally" check your spending, and your brain suddenly discovers a powerful new religion called *tomorrow*. Tomorrow you'll set up the budget. Tomorrow you'll call the insurance company. Tomorrow you'll look at the 401(k) portal you've been avoiding like it's cursed.

That's not laziness. That's biology. Willpower is not a moral virtue; it's a fuel source. It gets depleted by decision-making, stress, hunger, and plain old being awake. By night, most people are running on fumes, which is why so many financial promises are made in the morning and broken in the evening.

A habit-based money system does something radical: it stops asking your tired brain for permission. It treats good financial behavior the way a well-run building treats electricity. The lights come on because the wiring is right, not because someone stands by the switch all day "trying harder."

The difference between goals and systems (and why systems win)

A goal is a finish line. A system is a set of defaults that quietly steer you every week. Goals are inspiring; systems are dependable. Goals need motivation. Systems need design.

If you've ever set a goal like "save $5,000 this year" and then watched the months disappear, you've seen the trap. The goal lives in your calendar as a vague future tense. The system lives in your checking account on payday.

There's a reason the most effective savings strategy in the world is embarrassingly boring: automation. When money moves before you touch it, you don't have to be a disciplined person. You just have to be a person who set it up once on a reasonably good day.

A good money habit isn't heroic. It's *anti-heroic*. It's the small, repeatable action that keeps happening even when you're stressed, busy, or not feeling "motivated."

Habits form like ramps, not like staircases

People love the myth of the instant transformation. "Twenty-one days to a new habit." "Just decide." "Change your mindset." It sounds clean. Real habit formation is messier and more forgiving.

In real-world research on habit formation, automaticity tends to rise quickly at first, then slows and levels off. The early wins are real: the first few repetitions can make a behavior feel less awkward fast. But then you hit the part no one posts about: the plateau where it still takes effort, just less than before.

Even more important: the time it takes for a habit to feel automatic varies wildly by person and behavior. Some habits click in a few weeks. Others take months. That variance is not a sign you're broken; it's a sign you're human with a real life, a real schedule, and a nervous system that doesn't care about your spreadsheet.

Missing a day also isn't fatal. People tend to treat one slip as proof the whole plan doesn't work. The reality is closer to fitness: skipping one workout doesn't erase your muscles. It just means you do the next one. Consistency beats perfection because perfection is brittle.

So the target is not "never miss." The target is "never spiral." You want habits that bounce back.

Environment design: make the right thing the easy thing

If willpower is a battery, environment design is the wall outlet. You change outcomes by changing what's convenient.

Environment design means adjusting your surroundings—digital and physical—so that your default behavior is closer to what you actually want. It's not about self-control. It's about reducing the number of times you have to wrestle your own impulses.

A quick thought experiment: if your debit card disappeared for a week, what would happen to your spending? Not because you became wiser, but because spending would require extra steps. Friction is a tool.

The modern world is engineered to remove friction from spending. Saved cards, one-tap checkout, "buy now, pay later," app notifications timed for when you're tired. If you don't intentionally design your environment, you are living inside someone else's design—and their goal is not your net worth.

Here are practical environment moves that feel almost too simple until you try them:

- *Use two accounts on purpose.* One checking account for bills and "life," one for guilt-free spending. You're not hiding money; you're reducing daily decision fatigue.
- *Put saving on rails.* Automatic transfers scheduled to happen the day after payday (or the same day, if your cash flow allows it). Savings should happen before your brain gets a vote.
- *Make spending slightly annoying.* Remove shopping apps from your home screen, turn off retail notifications, and delete saved payment methods anywhere you tend to impulse-buy. Your future self will feel like you hired a bouncer for your wallet.

- *Name your savings accounts like they're real.* "Emergency Fund" is fine. "Rent Backup: 3 Months" is better. "Freedom Fund (Quit Bad Job)" is best. Your brain responds to meaning.

Environment design isn't a punishment. It's compassion for the version of you who makes decisions at 8:43 p.m.

Defaults: the quiet force that shapes your financial life

One of the most underrated facts in personal finance is that humans interpret defaults as recommendations. If a form comes with a box already checked, many people assume someone smarter than them set it for a reason.

Employers have used this to dramatically increase retirement plan participation through automatic enrollment. When saving is the default, participation jumps because inertia works in your favor. People who would never "get around to it" end up investing because the path of least resistance points toward their future.

But defaults have a dark side: people also tend to *stay* at default contribution rates and default investments for a long time. If the default is too low, inertia can create a slow-motion failure. The same force that helps you start can trap you at "good enough."

That means you should treat defaults like a draft, not a destiny.

If you have a workplace retirement plan:

- Check your contribution rate and make sure it reflects your goals, not your employer's minimum viable default.
- Learn what you're invested in. A default fund might be perfectly fine, or it might be overly conservative for your timeline.

If you don't have a workplace plan, you can still create defaults: automated transfers to savings, scheduled bill pay, auto-investing in a brokerage or IRA. You're building your own set of rails.

The mindset shift is powerful: instead of asking "How do I become disciplined?" you ask "What do I want to happen *by default*?"

A concrete case study: Maria, the nurse who stopped budgeting every day

Maria is a nurse in her early 30s. She's good at her job, exhausted by her job, and financially stuck in a way that looks confusing from the outside. She makes a solid income, but every month feels like it ends with the same magic trick: the money disappears.

Maria tried "being disciplined." She downloaded a budgeting app and checked it constantly. For about nine days, it worked. Then a string of night shifts hit, a family situation flared up, and her app became an icon she refused to look at. She described it like stepping on a scale after a week of junk food: the avoidance was the point.

What changed things wasn't a new category system. It was switching from daily willpower to a set of defaults:

- Her paycheck landed in checking on Friday.
- An automatic transfer moved money to a separate savings account on Saturday morning.
- Bills were on autopay from a "bills" checking account.
- She gave herself a weekly spending amount that moved every Monday into a separate debit account used for coffee, takeout, and fun.

Notice what's missing: daily tracking. Maria didn't become a spreadsheet person. She became a systems person.

The first month wasn't perfect, but it was different. She stopped bleeding money from dozens of micro-decisions because she reduced the number of times she had to decide. She also stopped feeling "bad with money" because she wasn't constantly failing a daily test.

A system that works at 30% motivation is a better system than one that requires 90% motivation. Maria didn't need to be a new person. She needed fewer decisions.

The Monthly Money Date: a ritual that replaces dread with clarity

Avoidance is expensive. It lets fees hide, subscriptions multiply, interest accumulate, and small problems become big ones. Most money stress is not caused by numbers. It's caused by *not knowing the numbers*.

The Monthly Money Date is a simple ritual: once a month, you sit down for 30–45 minutes and look at your financial life on purpose. Not daily, not obsessively. Monthly—like a checkup.

The rule is crucial: it must be judgment-free. The point isn't to scold yourself into better behavior. Shame doesn't create wealth; it creates hiding. You're not holding court. You're taking inventory.

Pick a consistent trigger: first Saturday morning, last Sunday evening, the day after payday, whatever fits your life. Put it on the calendar like a real appointment. The easiest way to keep it is to make it mildly pleasant: favorite drink, a specific playlist, a comfortable spot. You're training your brain to associate money with agency, not punishment.

Here's a practical script that works whether you're single or partnered:

Step 1: Open the "truth window" (10 minutes)

- Check your checking account balance(s).
- Check credit card balances.
- Check savings balance.
- If you invest, glance at contributions, not performance. You're verifying the machine is running.

The goal is not to feel good. The goal is to be informed.

Step 2: Find the leaks without drama (10–15 minutes)

Scan transactions and ask two questions:

- "What am I paying for that I don't value anymore?"
- "What surprised me?"

Surprise is your best diagnostic tool. It reveals either a missing category (you forgot annual expenses exist) or a behavior that's become automatic in the wrong direction (late-night delivery, convenience spending, subscription creep).

Cancel one thing. Just one. The compounding effect of one cancellation per month is shockingly large, and it builds the identity: *I pay attention.*

Step 3: Decide one adjustment (10 minutes)

Pick one small tweak that changes the next month:

- Increase an automatic transfer by $25.
- Set a spending limit for a problem category by moving that amount into a separate account.
- Turn on autopay to avoid interest.
- Schedule a call to negotiate a bill.

One adjustment keeps it from becoming a "review" ritual that changes nothing. Momentum comes from action.

Step 4: Close with a win (5 minutes)

Write down one thing that went right:

- "I paid every bill on time."
- "My credit card balance dropped."

- "I cooked at home three times."
- "I didn't avoid looking."

This matters because your brain learns through reward. If every money check-in ends with punishment, you will stop doing them. If it ends with a win—even a small one—you'll return.

If you share money with someone: turn it into teamwork

Money fights rarely start with math. They start with meaning: safety vs freedom, generosity vs control, "I deserve this" vs "we can't afford that." A Monthly Money Date with a partner works best when it's framed as collaboration, not interrogation.

A simple opening line can change the whole tone:

> "I'm not here to judge. I want us to feel calm about money. Can we look at the numbers and pick one improvement together?"

If the conversation gets tense, return to the shared goal: stability, options, less stress. Then focus on behaviors you can automate. Partners don't need identical money personalities. They need a system that protects both people from their worst financial instincts.

Make habits resilient: build for bad weeks, not perfect weeks

The biggest mistake in habit-building is designing a system that only works when life is smooth. Life is not smooth.

Resilient money habits have two features:

- *They're small enough to do when you're tired.*
- *They have a reset button.*

A reset button might be a rule like: "If I overspend, I don't 'get strict.' I move $50 less to fun money next week and keep everything else the same." No drama, no punishment spiral, just a small correction.

Another reset button is having a minimum version of the habit: "Even if I can't do the full Money Date, I will log in and check balances." That keeps the identity alive: *I'm someone who looks.*

You're not building a system for your best self. You're building it for the version of you that has a headache, a deadline, and three emotional texts waiting.

Actionable takeaway: design your ``one-hour money system"

Block one hour this week and do the following in order. Not someday. Pick a date and time.

1. *Create or rename accounts for clarity.* If possible: one bills account, one spending account, one savings account. Name savings with a purpose.
2. *Set one automatic transfer.* Even $25 per paycheck counts. The amount matters less than the existence of the rail.
3. *Turn on autopay for at least one bill or card.* Avoiding late fees is a high-return habit.

4. *Schedule your Monthly Money Date.* Put it on your calendar for the same day each month with a realistic duration (30–45 minutes).
5. *Add friction to one spending trigger.* Delete one saved card, disable one-click checkout, or remove one shopping app from your phone.

When you're done, you should feel a specific kind of relief: not the rush of a perfect plan, but the calm of knowing that some of your financial life will improve automatically while you're busy living it.

6.3 The Rent Versus Buy Dilemma

Someone will eventually say it at a barbecue, usually while holding a paper plate: *"Renting is throwing money away."* It lands like wisdom. It sounds adult. It also has the emotional subtlety of a foghorn.

That sentence survives because it's half-true in the laziest possible way. Yes, when you rent, you don't build home equity. But the hidden implication is the real problem: *when you buy, you stop "throwing money away."* That part is simply false. Buying has plenty of money you never get back. It's just wrapped in ceremony, paperwork, and the comforting smell of fresh paint.

A better question isn't "rent or buy?" It's: *What's the total cost of living in this home, in this city, for my likely timeline—and what else could my money do instead?* That's the grown-up version. It's less catchy, but it's how you keep a housing decision from quietly dictating the next decade of your finances.

The emotional trap: homes feel like safety, so we stop doing math

Housing sits at the intersection of money and identity. It's not only shelter. It's neighborhood, schools, commute, relationships, privacy, control, and status. That's why people can be ruthlessly analytical about their 401 (k) fees and completely unhinged about a granite countertop.

Your brain also loves the story of ownership. Owning feels like progress. Renting can feel like waiting. And if you grew up hearing "get a house" as the definition of making it, renting may carry a quiet shame even when the numbers scream that renting is the rational move.

The antidote isn't to become cold or cynical about homes. It's to separate the emotional value (stability, autonomy, pride) from the financial value (cost, risk, flexibility). Both matter. Confusing them is what creates regret.

The phrase that changes everything: *unrecoverable costs*

The cleanest way to cut through the noise is to focus on *unrecoverable costs*—money you pay to live somewhere and never see again.

Rent is unrecoverable. Everyone agrees on that.

But owning also has unrecoverable costs, and they are not small:

- *Mortgage interest:* the fee you pay the bank for using their money.
- *Property taxes:* a bill that doesn't build equity.

- *Homeowners insurance:* necessary, but not an "investment."
- *Maintenance and repairs:* the house will eventually demand tribute.
- *HOA dues (if applicable):* sometimes worth it, often maddening, always monthly.
- *Closing costs when you buy:* lender fees, title, escrow, and other "small" line items that add up fast.
- *Realtor commissions and selling costs:* the toll to exit.
- *Opportunity cost:* money tied up in a down payment and closing costs can't be used elsewhere.

Notice what's *not* on that list: principal payments. Principal is not a cost; it's converting cash into home equity. You can still argue about whether that equity grows or shrinks, but principal itself isn't "money thrown away." It's forced saving.

This is why the rent-versus-buy debate gets toxic when people compare rent to a mortgage payment. A mortgage payment is a cocktail. Some of it is unrecoverable (interest). Some is a bill (taxes and insurance). Some is a transfer to yourself (principal). Comparing the whole payment to rent is like comparing the cost of a restaurant meal to the cost of groceries without noticing that one includes the chef, the servers, and the vibe.

A picture worth a thousand arguments

The simplest way to see the issue is to stack the unrecoverable costs side by side. Over a five-year window, buying often carries a heavy front-loaded cost because of closing costs and early-year interest.

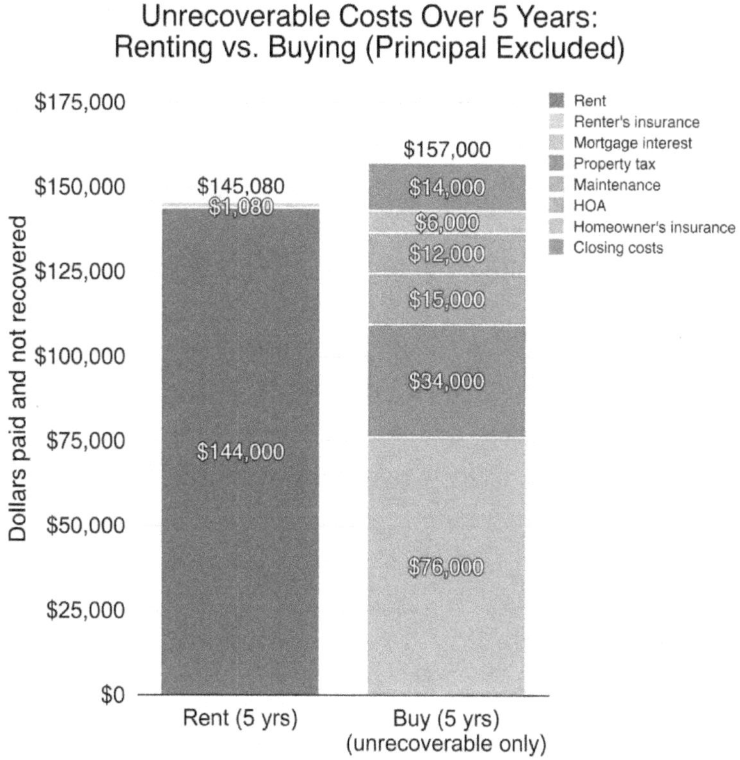

The point isn't that buying is "bad." The point is that buying is not a magical escape from unrecoverable costs. You still pay for housing. You just pay in different categories, with different timing, and different risks.

Total Cost of Ownership: the five buckets you have to count

To run a real comparison, you need a Total Cost of Ownership lens. Think in buckets, not slogans:

1) Transaction costs

Buying and selling are expensive events. Even if you never refinance, you'll pay meaningful one-time costs to get into and out of a home. These costs punish short timelines. If you might move in three years, transaction costs are the financial equivalent of sprinting in dress shoes.

2) Monthly carrying costs

Owning isn't just the mortgage. It's taxes, insurance, HOA, utilities that may be higher, and the steady drip of maintenance. People underestimate maintenance because it doesn't arrive on a schedule. It arrives when the water heater chooses violence.

3) Financing costs

Interest is the rent you pay to the bank. Early in the loan, interest is typically the dominant portion of your mortgage payment. That's why the "I'm building equity" feeling can be misleading in the early years: you're building *some* equity, but you're also paying a large interest bill.

4) Opportunity cost

A down payment isn't only money spent. It's money *committed*. You lose flexibility. You may also lose investment growth you could have earned elsewhere. Opportunity cost is the most ignored part of the rent-vs-buy conversation because it's invisible. Humans are terrible at valuing invisible alternatives.

5) Risk and uncertainty

Home prices can rise, stagnate, or fall. Neighborhoods change. Job markets change. Life changes. You don't control any of that, but housing can amplify it. Buying concentrates a big chunk of your net worth in one undiversified asset in one location, often financed with leverage.

Leverage is powerful in both directions. A 10% rise on a heavily leveraged purchase can feel like genius. A 10% drop can trap you, especially if you need to move and selling costs show up at the worst time.

A concrete example: two friends, one city, two timelines

Imagine two friends in the same metro area, both looking at a $450,000 starter home.

Ava is early in her career. She might change employers, maybe even cities, within the next three years. She's ambitious, and her income trajectory is upward but uncertain. She also values flexibility: the ability to say yes to an opportunity without having to sell a house first.

Ben is settled. He expects to stay put for at least seven to ten years. He cares about stability and wants control over his space. His job is relatively location-tied, and he's comfortable handling home maintenance (or paying for it without resentment).

If Ava buys and moves in three years, the math has a few sharp edges:

- Transaction costs get spread over a short period.
- Early mortgage payments are interest-heavy.
- A modest dip in home prices or a slow market can turn the move into a financial headache.

Even if Ava's home price rises, her *net* outcome depends on all the costs she paid to make that appreciation happen. She could win, but she's taking a bet she may not need to take.

Ben's timeline changes everything:

- Transaction costs get diluted over more years.
- He has more time for principal paydown and potential appreciation to matter.
- The non-financial benefits (control, stability) have more time to pay him back in quality of life.

Same home. Same city. Different correct answers, because their timelines and preferences change the total cost.

That's the part the barbecue slogan can't handle: housing is a math problem *and* a life problem.

``But my rent goes up." Yes. And your ownership costs do too.

Rent increases are real, and they're emotionally loud because you feel them as a direct hit.

Ownership costs inflate in quieter ways:

- Property taxes can rise as assessments change.

- Insurance premiums can rise, sometimes dramatically in certain regions.
- Maintenance costs inflate with labor and materials.
- HOA dues can creep upward.

Owning can provide payment stability on the mortgage *principal and interest* portion if you have a fixed-rate loan, which is meaningful. But "fixed" doesn't mean "frozen." The full housing payment can still change.

A calm way to think about it: renting exposes you to rent inflation; owning exposes you to a portfolio of housing-related inflation. Neither is immune. They're just different risk packages.

The non-financial benefits are real (and they deserve a price tag)

Owning can be worth it even if it's not the absolute best spreadsheet outcome. Stability, control, and the ability to customize your space are legitimate benefits.

The mistake is pretending those benefits are "free," then using them to justify a purchase that strains your budget. A healthier approach is to name the benefit and decide what it's worth.

For example:

> "I'm willing to pay an extra $300/month in total cost for stability and control, as long as my emergency fund stays intact and I'm still investing."

That's an adult trade-off. It respects both your emotions and your future.

Actionable takeaway: your rent-vs-buy checklist (no spreadsheets required to start)

Before you open a calculator, answer these questions in writing. The goal is to expose the assumptions that will drive the math.

1. *Timeline:* Realistically, how many years will I stay? If the answer is "I don't know," use a shorter timeline as your base case.
2. *Flexibility premium:* How valuable is the ability to move quickly for a job, relationship, or family need?
3. *Cash safety:* After down payment and closing costs, will I still have a meaningful emergency fund, or am I turning my safety net into drywall?
4. *Maintenance tolerance:* Am I willing to spend time coordinating repairs, or pay extra to outsource them without resentment?
5. *Opportunity cost honesty:* If I rent, will I actually invest the difference, or will it vanish into lifestyle upgrades?

Then run one simple test: calculate your five-year unrecoverable costs for both choices. If buying doesn't beat renting *by a wide margin* on your likely timeline, treat "renting is throwing money away" as what it is: a slogan that can't see your life. The best housing decision is the one that lets you sleep at night and still build wealth in the background, month after month.

6.4 Smart Mortgage Management

The mortgage office has a particular smell: coffee, printer toner, and your future quietly being decided in twelve-point font.

You sit across from someone friendly who has done this a thousand times. They slide a page toward you with a monthly payment circled in pen, like that number is the whole story. If you nod, the process continues. If you hesitate, the salesperson voice comes out: *"Homes are a great investment." "You can always refinance later." "This is what people do."*

Mortgages are where smart, capable adults accidentally over-leverage their lives. Not because they're reckless, but because the product is designed to look simpler than it is. A mortgage is not just a loan. It's a long-term contract that reaches into your next decade and rearranges what's possible: career moves, family decisions, stress levels, even sleep.

Managing a mortgage well isn't about becoming a rate-obsessed finance nerd. It's about understanding the mechanics well enough that you can't be hypnotized by a low monthly payment.

The anatomy of a mortgage payment (and the part everyone forgets)

A typical housing payment has multiple moving parts. The one most people focus on is *principal and interest*—the portion tied to the loan itself. But your actual out-the-door cost often includes property taxes and homeowners insurance (sometimes collected monthly through an escrow account). If the property has an HOA, add that too.

The dangerous mental trick is to treat the mortgage payment like rent: a fixed "housing bill." It isn't. Taxes and insurance can rise. Maintenance is real. And the loan payment itself hides a crucial detail: early on, most of what you pay is interest, not principal.

That's not a scam. It's arithmetic.

Why early payments are mostly interest (the front-loaded reality)

Mortgages in the U.S. are typically *amortizing* loans. "Amortizing" just means the payment is structured so the loan is paid off by the end of the term, with equal monthly payments (for fixed-rate loans). The payment stays the same, but the *mix* changes over time.

At the beginning, your balance is at its highest. Interest is calculated on that big balance, so the interest portion of your payment is big. As you slowly pay the balance down, interest shrinks, and more of your payment goes toward principal.

The result is counterintuitive: in the early years, it can feel like you're paying a lot and getting nowhere. That feeling is accurate.

Here's what that looks like visually.

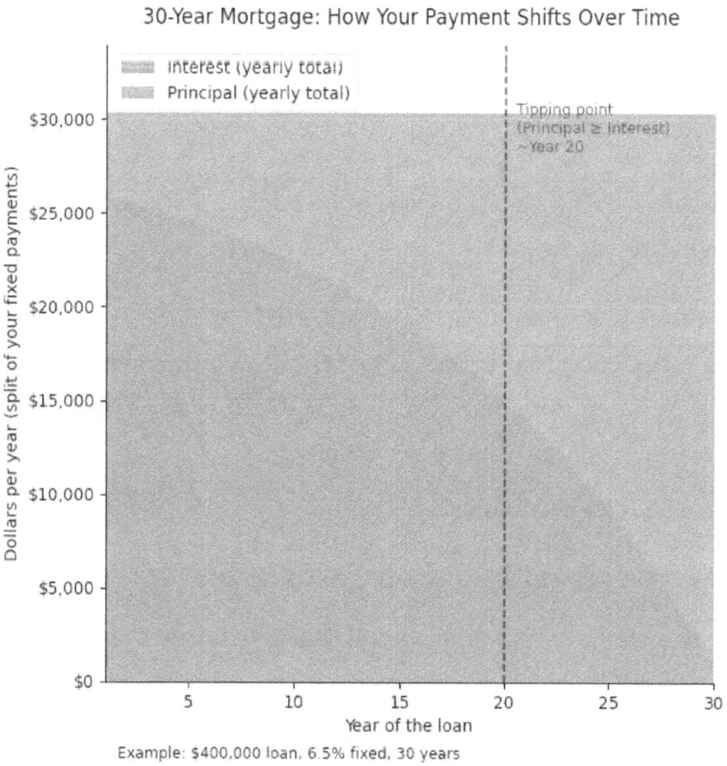

That curve explains two common homeowner emotions that people rarely admit out loud:

- *"I've paid for years and my balance barely moved."* Early amortization is slow on principal.
- *"Refinancing reset the clock."* When you refinance into a new 30-year loan, you often return to the interest-heavy early years (unless you choose a shorter term).

Understanding this doesn't mean you shouldn't take a 30-year mortgage. It means you should stop thinking of it as a neutral default. It's a specific choice with a specific cost.

The most expensive sentence in housing: ``We can afford the payment"

A monthly payment can be engineered. A lender can stretch the term, change the rate, adjust the down payment, include points, or propose an ARM. The payment is not a pure reflection of affordability; it's a variable.

Affordability is broader:

- Can you still save and invest consistently?
- Can you handle a job disruption without immediate panic?
- Can you pay for repairs without running to credit cards?
- Does the payment still work if taxes/insurance rise?

A mortgage that consumes your financial flexibility is not a *"dream."* It's a trap with nice countertops.

A useful rule of thumb: if buying the house requires you to be lucky for the next five years (no job loss, no major repairs, no life changes), the house is too expensive for your current life.

APR, points, and the break-even question (the part that actually matters)

When comparing loans, people fixate on the interest rate. Rate matters, but it's not the whole cost. Lenders can trade off rates and fees through *points* (upfront fees paid to reduce your rate) or *lender credits* (the reverse: a slightly higher rate in exchange for help covering closing costs).

That's why *APR* exists. APR (Annual Percentage Rate) is designed to summarize the cost of a loan by incorporating certain fees along with the interest rate, making it easier to compare offers with different combinations of rates and upfront costs.

The right way to evaluate points isn't "Is the rate lower?" It's "How long will I keep this loan?" Points only make sense if you stay long enough to break even.

Break-even is a simple concept:

$$\text{Break-even months} \approx \frac{\text{Upfront cost}}{\text{Monthly savings}}$$

If you pay $4,000 in points to save $80/month, the break-even is about 50 months. If there's a real chance you'll sell or refinance before then, you paid for a discount you didn't get to use.

That same logic applies to refinancing. Refinancing can be a smart move, but "You can always refinance later" is not a plan. It's a hope. A refinance only wins if your monthly savings and your time horizon beat the total cost to refinance.

Case study: Jamal almost bought the rate, then bought the timeline

Jamal is a first-time buyer with a stable job and a healthy down payment. Two lenders offer him the same loan amount.

Offer A: slightly higher rate, low upfront costs. *Offer B:* lower rate, but it requires points that add $5,000 to closing.

Jamal's first instinct is Offer B because it feels like "winning." A lower rate is a trophy. But Jamal is also likely to move in five years because his career track often relocates.

He does the unsexy calculation: the points save him $90/month. That's roughly 55 months to break even. If he moves at month 60, the "savings" are tiny. If he moves at month 48, the points were a donation.

So Jamal chooses Offer A. Not because it's universally better, but because it matches his timeline. He buys the loan that fits his life rather than the one that looks best on paper.

That's smart mortgage management: aligning the product with your reality, not your ego.

The 30-year vs 15-year question (and the third option people ignore)

A 15-year mortgage often comes with a lower interest rate and forces a much higher principal payment each month. The upside: you pay dramatically less interest over the life of the loan and you build equity fast. The downside: the monthly payment can strain cash flow and reduce flexibility.

The 30-year mortgage offers lower required payments and more breathing room. That breathing room can be used for investing, building an emergency fund, or simply not living on the edge. It can also be used for lifestyle inflation. The mortgage doesn't decide; your habits do.

There's a third option that many people ignore: take the 30-year mortgage (for flexibility) and then *voluntarily* accelerate it.

Two common acceleration strategies:

- *One extra payment per year.* If you make one additional principal-and-interest payment annually (or split it into small monthly additions), you can shave years off the loan and reduce total interest meaningfully.
- *Round up your payment.* If the payment is $2,463, paying $2,600 consistently makes a real difference over time.

Important detail: extra money needs to be applied to *principal* (and you should verify how your lender processes extra payments). Otherwise, you might be prepaying interest or simply paying ahead on the schedule without reducing the balance as intended.

This approach keeps flexibility. In a tight month, you pay the required payment. In a strong month, you attack principal.

The ``three business days" gift: use your review window

Mortgage closings feel like a sprint, which is convenient for everyone who benefits from you not noticing things. One of the strongest consumer protections in modern mortgages is the requirement that you receive the *Closing Disclosure* at least three business days before closing. That's not trivia. That's your leverage.

Those days are for reading like a professional skeptic:

- Compare the Closing Disclosure to the earlier Loan Estimate. What changed? Why?
- Scan for fees you don't recognize and ask for explanations in plain language.

- Confirm the interest rate, loan type, prepayment penalties (rare but not extinct), and whether anything is adjustable.

The goal is not to catch someone doing something evil. The goal is to catch the normal chaos of a complex transaction—the "small" fee errors, mismatched numbers, or last-minute surprises that happen when you're overwhelmed and eager to be done.

Shopping mortgages without getting gamed

Shopping matters. Even small changes in rate and fees can add up to tens of thousands of dollars over time. But "shopping" doesn't mean scrolling one website and trusting the rankings.

Some online comparison platforms are not neutral referees. They can be influenced by pay-to-play arrangements that shape what you see and what you don't. Treat any "best offer" list the way you'd treat a salesperson's "recommended" option: useful, but not authoritative.

A practical approach:

- Get multiple Loan Estimates from different lenders within a short window while rate shopping (so credit inquiries are treated appropriately).
- Compare APR and closing costs, not just the headline rate.
- Ask each lender to explain their fees and whether points or credits are included.

You're not being difficult. You're doing due diligence on the largest debt most people ever take on.

Actionable takeaway: the Mortgage Safety Checklist

Before you commit, answer these questions and do these actions. If you can't, you're not ready to sign—not because you're unworthy, but because the deal is bigger than vibes.

1. *My timeline:* Am I likely to keep this home (and loan) for at least 5–7 years? If not, I treat points and heavy closing costs as suspicious.
2. *My buffer:* After closing, will I still have an emergency fund? If the answer is no, the down payment is too large or the home is too expensive.
3. *My true housing cost:* Can I afford taxes, insurance, HOA, and a maintenance reserve *without* reducing retirement contributions to zero?
4. *My comparison:* I have at least two Loan Estimates to compare side by side.
5. *My review:* When I receive the Closing Disclosure, I will compare it to the Loan Estimate line by line and ask questions during the three-business-day window.
6. *My leverage plan:* If I choose a 30-year mortgage, I commit to a specific principal-acceleration habit (extra payment, rounding up, or a fixed extra amount monthly) that I can sustain without stress.

A mortgage can be a tool that stabilizes your life and builds long-term wealth. It can also be a beautifully packaged financial chokehold. The difference is rarely intelligence. It's whether you understand the mechanics well enough to insist on a loan that fits your timeline, your cash flow, and your ability to breathe.

6.5 Growing Your Income Engine

If you're trying to build wealth from scratch, there's a moment when extreme frugality turns from "responsible" to "small." Not morally small—strategically small. You can only cancel the same subscription once. You can only optimize groceries so many times before you're just arguing with yourself in aisle seven about the ethics of name-brand pasta.

Income is different. Expenses have a floor. Income has a ceiling you haven't met yet.

That doesn't mean you should spend wildly. It means the biggest lever in your entire financial life usually isn't a coupon. It's your earning power—what you get paid for your time, your skills, your judgment, and your ability to solve annoying problems other people avoid.

Most people treat their income like the weather: something that happens to them. Smart people treat it like an engine: something you maintain, upgrade, and occasionally rebuild so it can carry heavier loads.

Why your paycheck matters more than your budget

A budget is essential because it creates margin—the gap between what you earn and what you spend. Margin is where saving, investing, and debt payoff live.

But there's a hard truth hiding in the math: when margin is tiny, the budget becomes a knife fight. Every "unexpected" expense is a crisis. Every choice is trade-offs you can feel in your chest. And because the stakes feel high, it's easy to burn out and quit.

Raise income and the same discipline suddenly produces dramatically more results. An extra $500/month of take-home pay, invested or used to pay down high-interest debt, can change the trajectory of a decade. Not in a motivational-poster way. In a literal "this is now possible" way: emergency fund, retirement contributions, better housing options, time back, stress down.

Income also reduces fragility. A small emergency is only an emergency when you don't have slack.

The real income skill: becoming harder to replace

Employers don't pay you based on your effort. They pay you based on your *replaceability*. If you can be swapped out with minimal cost, your pay will tend to drift toward the minimum they can get away with. If replacing you is painful—because you have rare skills, deep context, strong relationships, or you reliably fix expensive problems—your leverage improves.

This is why "work hard" advice often fails. You can work hard at the wrong things and become a perfectly replaceable expert in busywork. The goal is to build *career capital*: valuable skills and credibility that compound.

Career capital comes from:

- Skills that map to revenue, cost reduction, risk reduction, or time savings
- Work that is visible (or at least legible) to decision-makers
- Proof—numbers, projects, outcomes—that you can point to without blushing

The key is to stop thinking of your job as your identity and start thinking of it as your current platform for acquiring leverage.

Present bias is why most people underinvest in themselves

If buying a new phone gives an immediate dopamine hit, learning a new skill often gives immediate discomfort. You feel clumsy. You confront gaps. You sit with the quiet humiliation of being a beginner again.

That's present bias at work: your brain discounts future rewards and overweights present pain. It whispers, *"Not now. You're tired. You can learn later."*

Later becomes a year. Then three.

The people who grow their income consistently aren't necessarily more talented. They're better at making future rewards feel urgent and making present effort feel automatic. They don't rely on motivation. They rely on defaults: recurring calendar blocks, learning routines, and social accountability that makes quitting inconvenient.

If your career growth depends on "when you have time," it will be starved to death by life.

A case study: Danielle's $18,000 raise without changing companies

Danielle works in operations at a mid-sized company. She's been "the reliable one" for years: always helpful, always available, always putting out fires. Her reviews are glowing. Her pay is not.

When Danielle finally decides to ask for a raise, she does what most people do at first: she prepares an emotional argument. She lists how hard she works. She mentions loyalty. She explains that everything is more expensive now. It's honest, and it's also the language of someone asking for kindness.

Then she shifts to leverage.

For one month, Danielle keeps a simple log. Nothing fancy: a note on her phone with dates and outcomes. She captures three categories:

- Problems she prevented (errors caught before they shipped, vendor issues resolved early)
- Time saved (process changes, automation, templates, documentation)
- Money saved (avoided penalties, reduced waste, improved inventory accuracy)

She also does one strategic thing: she picks a high-visibility problem that annoys leaders and solves it. Not heroically—systematically. She documents the before-and-after and makes it easy to understand.

When she goes into her compensation conversation, she doesn't plead. She positions.

She says, calmly: the scope of her role has grown; she's producing measurable value; and she wants her compensation to match the market and her impact. She brings a specific number. She also brings an option: a raise now, or a defined promotion path with a timeline and metrics.

Her manager doesn't instantly approve it. But Danielle has changed the game. The conversation becomes about business value, not feelings. Two months later, she gets an $18,000 raise and a title adjustment.

Danielle didn't suddenly become a different person. She learned to speak the language money responds to: outcomes.

Salary negotiation: the scripts that keep you calm

Negotiation is a skill, not a personality trait. You don't need to be aggressive. You need to be clear. The goal is to create a business conversation where the other person can say yes without losing face.

A few scripts that work because they're direct and emotionally neutral:

When you want a raise

"I'd like to talk about my compensation. Over the last __ months, my responsibilities have expanded in these ways: __. Here are outcomes I've delivered: __. Based on the scope of my role and market rates, I'm targeting $__. What would it take to make that happen?"

When they say "not right now"

"I understand. If we can't adjust compensation today, can we define a clear path? What specific targets would you need me to hit, and by what date, to revisit this?"

Then follow up in writing. Vagueness is where raises go to die.

When you're negotiating a new offer

"I'm excited about the role. If we can get to $__ base (or $__ total comp), I'm ready to sign."

Then stop talking. Silence is not awkward; it's leverage. People often fill silence by negotiating against themselves.

When you're worried you'll sound ungrateful

Gratitude and negotiation can coexist: *"I appreciate the offer and I'm genuinely excited. I also want to make sure the compensation reflects the scope of the role and the value I'll be bringing."*

This keeps the tone warm while still holding the line.

Skills ROI: don't ``learn more,'' learn *profitably*

The internet will happily sell you a new certification every week. Most won't change your income. They'll change your resume in a way that feels productive while producing very little.

Skills ROI means treating skill-building like an investment:

$$\text{Skills ROI} = \frac{\text{Expected increase in annual earnings}}{\text{Total cost (money + time + stress)}}$$

You don't need perfect math. You need disciplined questions.

Before you pay for a course, ask:

- *Does this skill show up in job postings that pay more?* Not in "nice to have" bullets—in the requirements and responsibilities.
- *Can I build a portfolio project or proof of work with it?* Skills without evidence don't travel.
- *Is there a cheaper way to validate the path?* A free tutorial, a small project, informational interviews, or a part-time task at work that uses the skill.
- *Does it fit my current leverage?* A skill is more valuable when it stacks on top of what you already do well.

High-ROI skills tend to share a trait: they make you useful across companies. Things like data analysis, project management, sales ability, writing clearly, automation, and managing stakeholders. Even in technical fields, the rarest skill is often the ability to translate complexity into action without making everyone hate you.

Upgrade your network without becoming a ``networking person"

Networking has a branding problem. People picture forced small talk and business-card rituals. Real networking is simpler: it's being known as competent and easy to work with, then keeping light contact with people as you all move through your careers.

A practical, low-cringe approach:

- Once a week, send one message to someone you respect: a quick check-in, a thank-you, or a "saw this and thought of you."
- Once a month, have one coffee chat or virtual chat with someone one step ahead of you.
- When you finish a good project, write a short post or internal note about what you learned. Not bragging—proof of work.

This creates optionality. Optionality is the hidden asset that makes negotiating power real.

Lifestyle inflation: the silent killer of raises

The most common failure mode after increasing income is not bad investing. It's upgrading your entire life at the same speed as your paycheck.

Lifestyle inflation isn't inherently evil. If more money means a safer neighborhood, childcare support, therapy, or time back, that can be a wise trade. The danger is unconscious upgrades: the nicer car "because you can," the daily convenience spending, the bigger apartment because it looks like progress, the recurring monthly commitments that turn a raise into a new baseline.

Raises are most powerful when you split them:

- Some goes to improving life now (so the plan is sustainable)
- A meaningful portion goes to future wealth (so the plan actually works)

A simple rule that prevents accidental sabotage: whenever your income rises, increase your automatic investing/saving the same week. Not later. Later is where money goes to disappear.

If you don't capture the raise intentionally, your brain will spend it for you in a hundred small ways that feel like "nothing."

Build ``career defaults" the way you build money defaults

Earlier, the idea of defaults mattered because inertia is powerful. The same principle applies to your income engine. You can design career growth so it happens even when you're busy.

Examples of career defaults:

- A recurring weekly calendar block for skill-building (protected like a meeting)
- A monthly review of your accomplishments log (so you don't rely on memory during negotiation)
- A quarterly "market check" where you scan job postings to see what skills are being paid for
- A rule that you apply to a certain number of roles per month when you're underpaid, even if you like your job

This doesn't mean you're disloyal. It means you refuse to let inertia price your labor.

Actionable takeaway: the 30-day income upgrade plan

Set a 30-day timer and run this plan like a small project. Not a life overhaul—a focused sprint.

1. *Start an accomplishments log today.* Create a note titled "Proof of Value." Each week, add 3–5 bullets: what you shipped, what you improved, what you prevented, what you saved (time/money/risk).
2. *Choose one high-ROI skill to build.* Pick the skill that appears repeatedly in higher-paying roles you'd actually want. Commit to two 45-minute blocks per week on your calendar.
3. *Have one leverage conversation.* Either ask for expanded responsibility that builds a portfolio outcome, or ask your manager directly: *"What would make me the obvious choice for a raise or promotion?"*
4. *Do one market reality check.* Apply to 3 roles (or do 3 informational chats) even if you're not desperate to leave. You're testing your market value, not begging for rescue.
5. *Pre-commit your next raise.* Decide, in writing, that when income increases, a fixed percentage will automatically go to investing, debt payoff, or your emergency fund before lifestyle upgrades.

The income engine doesn't roar to life because you suddenly feel confident. It grows because you build proof, stack skills that pay, and make career progress the default setting rather than an occasional burst of motivation.

6.6 Evaluating Side Hustles

At some point, almost everyone has the same late-night thought: *"If I could just make an extra $500 a month, everything would calm down."* The car repair wouldn't turn into a crisis. The credit card balance would finally shrink. You could breathe without doing mental math at the grocery store.

Then the algorithm smells hope and serves you an all-you-can-eat buffet of side hustle fantasies. Drive your car! Rent your spare room! Flip sneakers! Start a "passive income" store while you sleep! The pitch is always the same: *easy money, fast.* The reality is always the same too: most side hustles are either low-paid labor with a shiny name, or real businesses disguised as hobbies that require far more skill than the ad implies.

A side hustle can be a powerful tool. It can also be a clever way to exhaust yourself while earning less than you think. The difference comes down to one word: *unit economics.*

Unit economics: the only question that matters

Unit economics means you measure what you earn *per unit* of effort after *all* costs. The "unit" might be an hour, a ride, a delivery, a product sold, a client project. If the unit economics are bad, scaling just makes you tired faster.

Most people accidentally do side hustle math like this:

$$\text{Profit} = \text{Money that hit my account}$$

That's not profit. That's revenue. Revenue is the satisfying part. Profit is the truthful part.

Real profit looks more like:

Profit = Revenue – platform fees – expenses – taxes – your time

Yes, you subtract your time. Not because time is "priceless," but because time is scarce. A side hustle that pays $18/hour net might still be fine if your alternative is scrolling. It's not fine if it prevents you from finishing a certification that would raise your salary by $15,000 a year.

The hidden costs that eat side hustles alive

Side hustles are notorious for costs that hide in plain sight. A few that show up again and again:

Platform fees and pricing control

Apps and marketplaces can take meaningful cuts. They can also change the rules, change the algorithm, and flood the market with new workers. You don't control pricing; you rent access.

Transaction costs

Driving to pick up supplies. Messaging leads who ghost you. Cleaning between guests. Packaging and shipping. These are real hours that don't show up as "work," but they absolutely show up in your fatigue.

Equipment and depreciation

If you drive for money, your car is a business asset. Assets wear out. Tires, brakes, oil changes, unexpected repairs, and the slow decline in resale value are part of the cost of each dollar you earn. If you ignore depreciation, you can convince yourself you're profitable while quietly consuming your car.

Taxes (the surprise bill)

Side hustle income is generally taxable. If you're an independent contractor, taxes aren't withheld automatically the way they are at a W-2 job. That means it's on you to set money aside, and in many cases to pay estimated taxes during the year.

People get burned here because the money feels like a bonus. Then tax time arrives like a bouncer: *"Great. Now pay up."*

Even if you receive no tax forms, income can still be reportable. The government does not accept "nobody told me" as a strategy.

Benefit loss and burnout

If the hustle pushes you to reduce hours at your main job or damages your performance, the true cost can be enormous. The highest-paying thing you do is usually the work that protects your primary income engine. Any hustle that undermines it is a financial own-goal.

A concrete case study: rideshare driving that looked profitable until it wasn't

Kevin decides to drive rideshare on weekends. In his first month, he's thrilled: $1,200 hits his account. He tells a friend, *"This is easy money."* He starts imagining how fast he'll pay off debt.

Then the real math shows up.

Kevin drove 900 miles that month. His car's operating costs are not just gas. They include maintenance and depreciation. Even using conservative numbers, the cost per mile adds up quickly. Add in platform fees and the reality that some hours were dead time waiting for rides, and the net hourly rate shrinks.

Then there's tax. Kevin is not used to setting aside money for self-employment taxes and income taxes. By April, he owes a chunk he didn't plan for, and he ends up putting part of it on a credit card. The hustle that was supposed to reduce debt briefly increases it.

Is rideshare *always* bad? No. For some people, in some markets, at some times, it can be a useful cash-flow patch. But Kevin's lesson is the one most people learn late: if you don't measure net profit per hour, you're doing motivational math, not business.

The side hustle spectrum: cash now vs wealth later

Side hustles come in different flavors. They're not all competing in the same sport.

Cash-now hustles (time-for-money)

These are gigs where income is closely tied to hours worked: driving, delivery, task services, shift work, many forms of freelancing when you're starting out. The advantage is immediacy. The downside is the ceiling: you can't scale without adding hours, and hours run out.

Cash-now hustles can still be worth it when:

- You need short-term margin to stop a financial bleed (high-interest debt, overdue bills)
- The work is flexible and doesn't hurt your main job
- The net hourly pay is actually decent after expenses and taxes

Asset-building hustles (time-now for leverage later)

These are slower, often more frustrating at first, and much more powerful over time: building a small service business with repeat clients, consulting, tutoring with a specialty, creating a product, building software, content that leads to clients, a niche e-commerce brand with real differentiation.

The early days can look "worse" on the spreadsheet because you're investing time before results. That's where present bias tries to sabotage you. Your brain prefers the gig that pays tonight over the skill that pays next year.

But leverage is how side hustles become wealth engines instead of exhaustion projects.

A simple filter: the Side Hustle Scorecard

Before you commit, run every idea through these questions. They're designed to protect you from the two classic mistakes: underestimating costs and overestimating scalability.

1) What is my true net hourly rate?

Estimate conservatively:

$$\text{Net hourly} \approx \frac{\text{Revenue} - \text{fees} - \text{expenses} - \text{tax set-aside}}{\text{total hours (including admin)}}$$

If you can't estimate it, you're not ready to scale it.

2) Does this hustle build a skill that increases my main income too?

The best hustles have a double benefit. For example, tutoring can build communication skills and subject expertise. Freelance analytics can build portfolio projects that lead to a higher-paying job.

If a hustle earns money but teaches you nothing transferable, it's fine as a short-term patch—but treat it like one.

3) Is there a ceiling? Who sets it?

If the platform sets prices and can replace you easily, the ceiling is usually low. If you control pricing through a differentiated skill and direct client relationships, the ceiling is higher.

4) Will this still feel worth it at hour 200?

Almost anything sounds tolerable at hour 10. Hour 200 is where reality lives. If the hustle requires constant customer drama, unpredictable scheduling, or heavy physical strain, it may collapse under its own weight.

5) What risk am I taking?

Financial risk (buying inventory, paying for ads), legal risk (working without proper licensing), and scam risk. A legitimate opportunity does not require you to pay to "unlock" the job. Any hustle that involves upfront fees, urgency, or complicated payment schemes deserves extreme suspicion.

Taxes and recordkeeping: the unglamorous difference between a hustle and a mess

If you earn side income as an independent contractor, the IRS generally expects you to keep records and report income. This is not optional. The fastest way for a side hustle to become a stress machine is to treat taxes like a future problem.

A few practical habits keep you safe without turning you into an accountant:

- *Separate your money.* Use a separate checking account for side hustle income and expenses if possible. Clean separation is the cheapest form of sanity.
- *Set aside tax money automatically.* Many people choose a flat percentage of each payment (for example, 20–30%

depending on income level and state) into a "tax" savings bucket. The exact percent varies, but the habit is the point.

- *Track expenses as you go.* Mileage logs, supplies, software, home office expenses (if applicable), fees. Don't rely on memory.
- *Be aware of estimated taxes.* If you earn enough without withholding, you may need to pay quarterly estimated taxes or adjust withholding at your day job to cover it. The goal is to avoid a nasty surprise bill and potential penalties.

None of this is exciting. That's why it works. Most people don't do it.

How to avoid side hustle scams without becoming paranoid

The scam economy feeds on two emotions: urgency and shame. Urgency makes you act before you think. Shame makes you hide what happened.

A few red flags that deserve a hard stop:

- You have to pay money to get a job (training fees, starter kits, "background checks" through their link)
- They send a check and ask you to send some money back or buy equipment from a "partner"
- The offer is vague about the actual work but specific about the money
- They pressure you to act immediately or keep it secret

Real work is usually boring at the beginning. Real clients ask questions. Real businesses don't need you to wire money to "activate" your earnings.

Choosing the right hustle for your season of life

There's a reason side hustle advice often fails: it ignores seasons. The right hustle for a single person with evenings free is different from the right hustle for a parent with unpredictable childcare. The right hustle for someone paying off high-interest debt is different from the right hustle for someone already stable and trying to build long-term optionality.

Ask yourself what you need most:

- *Immediate cash flow?* Choose the highest net hourly option you can sustain without damaging your main job.
- *A skill upgrade?* Choose a hustle that forces you to practice a valuable skill with real stakes.
- *A scalable business?* Choose something with direct customer relationships and pricing power, even if the first month pays less.

The mistake is picking a hustle for the fantasy version of you. Choose for the version of you that exists on a Tuesday.

Actionable takeaway: the 14-day side hustle experiment

Don't "start a side hustle." Run an experiment. Two weeks is long enough to get real data and short enough to avoid sinking months into a bad idea.

Here's the experiment:

1. *Pick one hustle idea and define the unit.* Example: "one tutoring session," "one delivery shift," "one logo project," "one Etsy product shipped."
2. *Track everything for 14 days.* Revenue, fees, direct expenses, miles, and *total time* (including setup, messaging, travel, and admin).
3. *Set aside tax money immediately.* Choose a conservative percentage and move it to a separate bucket the same day you get paid.
4. *Calculate net hourly rate.* If it's lower than you expected, don't argue with the math. Adjust or quit.
5. *Make one leverage move.* Ask: "How do I raise my rate or reduce my time per unit?" That might mean specializing, improving your process, or moving off-platform to direct clients where appropriate and allowed.

At the end of 14 days, you'll have something most people never get: the truth. The truth lets you choose a hustle that actually helps—one that pays you fairly, fits your life, and doesn't quietly convert your evenings into low-wage labor with a motivational hashtag attached.

Chapter 7

Advanced Planning and Protection

You've worked hard to build your financial castle; now it's time to dig the moat. Many people assume that once they have money, the hard work is done, but wealth brings its own set of fragile complexities. In this chapter, we stop playing offense and start playing defense. We'll look at how to shield your life's work from lawsuits, market crashes, and health crises, and ensuring that your money ends up exactly where you want it to go when you're no longer here to guide it.

7.1 Insuring Against Catastrophe

Insurance gets sold like a vitamin: *Take this and you'll be healthier.* In real life, it's closer to a fire extinguisher. You don't buy it because you expect to use it; you buy it because if you ever *need* it and don't have it, your finances can go from "fine" to "ruined" in a single afternoon.

That framing is the difference between being an easy customer and being a hard one. Easy customers insure everything because the pitch feels comforting. Hard customers insure the few risks that can *wipe them out* and pay for the annoyances themselves. Hard customers don't treat insurance as an investment, a savings plan, or a membership program. They treat it as what it is: a contract to transfer specific, ugly risks you cannot afford to keep.

The reason people overbuy the wrong kind of insurance is emotional math. A cracked phone screen hurts your mood; a lawsuit or a house fire can hurt your entire life. Yet the phone screen is vivid (you see it daily) and catastrophe is abstract (until it isn't). Insurance companies understand this perfectly. They're happy to sell you peace of mind on small, frequent losses because it's profitable. Your job is to spend your insurance dollars where they buy *survival*, not vibes.

The real purpose of insurance: moving a boulder off your chest

Picture your finances as a person carrying a backpack. Some "risks" are pebbles: irritating, but you can keep walking. Some are boulders: one slip and you're pinned. Insurance is how you pay someone else to carry the boulder *for you*—with a catch. They won't carry every boulder. They carry the boulders you specify, in the ways you specify, under the conditions you specify. That means the real skill isn't "having insurance." The skill is *buying the right contract*.

A good insurance decision starts with one blunt question:

> If this happens tomorrow, can I pay for it without changing my life?

If the answer is yes, you're looking at an inconvenience. If the answer is no—if it would force debt, selling investments at a bad time, pulling a kid from a school, moving, or risking bankruptcy—you're looking at a catastrophe. Catastrophes are what insurance is for.

Right after you say "transfer the catastrophe," you need a simple way to sort everything else. Here's the mental model that keeps you from paying Cadillac premiums to cover bicycle problems.

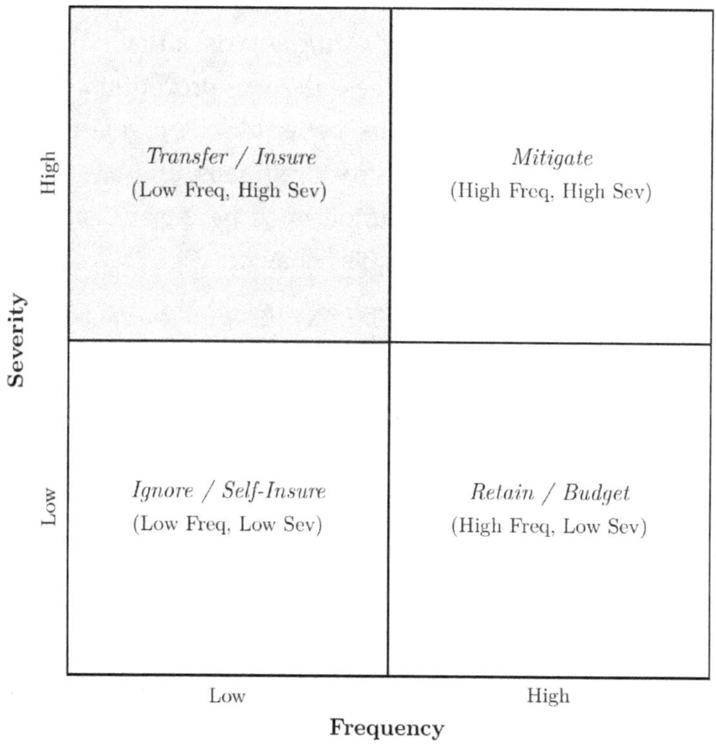

The beauty of this matrix is how quickly it reveals where your money should go.

High frequency / low severity (Retain/Budget): the stuff that happens all the time and costs a few hundred bucks. Car tires. Small dental work. A broken appliance. If you insure these heavily, your premiums become a disguised subscription for predictable expenses—with overhead and profit baked in. Better plan: keep a buffer in cash and handle them calmly.

Low frequency / low severity (Ignore/Self-Insure): the one-off annoyances that are unlikely and cheap. You don't need a policy for every weird scenario a salesperson can imagine.

High frequency / high severity (Mitigate): risks that happen often and would hurt badly. These deserve prevention, not just insurance: safer driving habits, better locks, cybersecurity hygiene, replacing a failing roof before it fails, and basic health maintenance. Mitigation is underrated because it isn't a bill you can pay to feel virtuous; it's behavior.

Low frequency / high severity (Transfer/Insure): the catastrophes. This is where insurance shines: house fires, major liability claims, severe storm damage, serious medical events, a totaled car that you rely on to earn income. These are rare, but the severity is life-altering.

A case study: the water that isn't ``covered"

A real-world pattern shows up after almost every major storm: two neighbors, same rainfall, wildly different financial outcomes.

Homeowner A has a standard homeowners policy and assumes "water damage" is water damage. Homeowner B has a standard policy too, but also has separate flood coverage because they learned the uncomfortable truth: homeowners insurance commonly excludes flood. When water comes from outside and inundates the home—rising water, storm surge, overflowing bodies of water—it's typically treated as flood, not the kind of accidental interior water damage people imagine when they hear the phrase.

Now the storm hits. Water pushes in through the door and seeps up through the foundation. Floors buckle, drywall wicks moisture like a sponge, cabinets swell, electrical systems need inspection, and mold becomes the silent second disaster. Homeowner A files a claim and learns that the biggest cost is excluded. The family drains savings, then reaches for credit, then delays repairs, then lives in a half-functional space while still paying a mortgage. Homeowner B files a claim and spends the next months dealing with contractors and paperwork too—but without the same financial free-fall.

That gap is what catastrophe planning is about. It's not about buying *more* insurance. It's about buying the *right* insurance for the risks that can permanently damage your balance sheet.

The uncomfortable part: many insurance disasters aren't caused by the event itself. They're caused by assumptions. People assume their landlord's policy covers their stuff (it usually doesn't). They assume homeowners covers flood (often excluded). They assume their auto policy covers every situation (coverage depends on the cause of loss, endorsements, exclusions, and limits). They assume "full coverage" is a thing (it isn't a standardized promise; it's a marketing phrase).

The deductible move: the simplest way to stop overpaying

There's a lever inside most insurance policies that quietly determines whether you're buying catastrophe protection or buying a pricey convenience plan: the *deductible*.

The deductible is the portion you agree to pay before the insurer pays. Raising your deductible often reduces your premium, and it does so in the healthiest possible way: it pushes small claims back onto you (where they belong) while keeping the insurer on the hook for big claims (where *they* belong).

Think of a deductible as a built-in filter. A $250 deductible invites frequent claims. A $1,000 or $2,500 deductible makes you pause and ask, "Is this worth filing?" That pause is valuable. Small claims can have hidden costs: potential rate increases, claim history that follows you, time and stress, and the temptation to treat insurance like a maintenance plan.

A good target is a deductible you can pay from cash without drama. Not from a credit card you'll carry for months. Not from selling investments in a downturn. Cash. If your emergency fund is thin, you might choose a lower deductible temporarily. But the long-term direction for a wealth builder is clear: *use insurance for disasters, not dents.*

Insurance math that matters (and the math that doesn't)

Most people judge insurance by one question: "Will I get my money back?" That's the wrong yardstick. A fire extinguisher isn't a bad purchase because it didn't get used. Catastrophe insurance is the same.

Here's a better way to think about it: you're paying a known, manageable cost (premium) to avoid an unknown, unmanageable cost (catastrophic loss). That trade is inherently "unfair" in a narrow sense because the insurer needs to pay claims, cover overhead, and earn a profit. Your goal is not to outsmart the insurer on expected value. Your goal is to *buy stability.*

What you *can* control is whether you're paying for risk transfer or for expensive noise:

- Policies with low deductibles and lots of extras tend to cover frequent, small losses. That's the most expensive way to use insurance.
- Policies with higher deductibles and strong liability/property limits tend to cover low-frequency, high-severity losses. That's the most financially sane way to use insurance.

If you want a surprisingly powerful financial habit, it's this: whenever you're quoted a policy, ask for the premium at three different deductibles. You'll see the price of convenience in dollars, not in vibes.

The coverage gaps that quietly wreck people

Catastrophe planning is mostly about spotting *gaps*. The biggest financial hits often come from areas where a person is "insured" in the casual sense but not protected in the specific, contractual sense.

Flood and earthquake: Many people learn the exclusions only after the event. If you live anywhere near water—and "near" includes places that have never flooded in your lifetime—it's worth exploring separate flood coverage, even if you're not in a high-risk zone on a map. Earthquake is similar: often a separate policy or endorsement, depending on where you live.

Renters: Renting doesn't make you immune to catastrophe; it just changes the shape. Your landlord's insurance is for the building. Your stuff, your temporary living costs after a covered loss, and your liability are typically your responsibility. Renters insurance can be surprisingly inexpensive because the insurer isn't insuring the building—but it can still be the difference between "annoying week" and "financial faceplant."

Liability limits that are too low: People obsess over insuring their phone and underinsure the risk that can garnish their wages. Liability is where normal life can turn into court dates: a car accident with injuries, a guest who slips and suffers a serious injury, your dog biting someone, a social-media-related claim in extreme cases. If your liability limit is small, a large claim doesn't shrink to fit it. It spills over into your life.

Replacement cost versus "depreciated" payouts: For property coverage, the difference between replacement cost and actual cash value can be the difference between rebuilding your life and getting a check that feels like an insult. Depreciation is brutal when you're trying to replace a household full of "old" items at today's prices.

A practical way to choose what to insure

You don't need to become an insurance nerd. You need a repeatable decision process that keeps you from buying what's shiny and skipping what's essential.

Start with your personal catastrophe list. For most households, it looks like this:

- *Being sued for more than you can pay* (liability).
- *Losing your home or suffering major property damage* (dwelling, plus the cost of living elsewhere during repairs).

- *Losing the use of your car* if you depend on it for work or family logistics (auto physical damage, plus rental coverage depending on your setup).
- *A medical event that creates large out-of-pocket costs* (health insurance choices do this job, but the catastrophe mindset applies: protect against the big hit).

Then apply the matrix:

1. Put every "insurance temptation" into one of the four quadrants.
2. For anything in *Retain/Budget*, increase your cash buffer instead of buying more coverage.
3. For anything in *Mitigate*, spend time and effort before spending premium dollars. Prevention has a better return than paperwork.
4. For anything in *Transfer/Insure*, buy high-quality coverage with limits that match the disaster, not the minimum a website defaults to.

This is where people ask, "So should I insure my phone?" The matrix usually answers: high frequency, low severity. Budget for repairs or replacements. If losing the phone would create a true catastrophe (you can't work, you're in a safety situation, you have no cash buffer), that's less an insurance problem and more a cashflow and emergency fund problem.

How insurers price your behavior (and what to do about it)

One more counterintuitive truth: the cheapest-looking insurance decision can be expensive over time.

Insurers price the probability you'll file claims. If you file small claims frequently, you look like a high-cost customer. Even if each claim is legitimate, the pattern matters. That's another reason higher deductibles and self-insuring small losses are powerful: you keep insurance in its proper lane, and you protect your future pricing.

There's also a behavioral benefit. When you stop expecting insurance to smooth out every bump, you start building the financial muscles that actually create wealth: maintaining a cash cushion, doing basic preventative maintenance, and making deliberate tradeoffs instead of panicked ones.

The goal: be boringly resilient

The ideal insurance setup is not the one with the most features. It's the one that makes your financial life boring after something terrible happens. Not painless. Not fun. Just *survivable.*

If a tree crushes your roof, you want the stress to be contractors and inconvenience—not whether you'll lose the house. If you cause an accident, you want the stress to be recovery and responsibility—not whether a lawsuit will follow you for a decade. If a storm floods your neighborhood, you want the stress to be cleanup and disruption—not the sinking realization that the biggest line item is excluded.

Actionable takeaway: your 30-minute catastrophe audit

Set a timer for 30 minutes and do this once a year:

1. Write down the three biggest financial disasters you could face in the next 12 months (not the weird ones, the plausible ones).
2. For each, answer: *Do I have a policy that specifically covers this? What is the deductible? What is the limit?*
3. Raise deductibles to the highest level you can comfortably pay from cash, and redirect the premium savings into an emergency fund until that deductible is truly "no big deal."
4. Skim for exclusions that match your actual life (flood, earthquake, water backup, high-value items, business use of a vehicle, etc.) and ask one direct question: *What would make this claim denied?*

If you can look at your insurance and say, "Small stuff is on me; big stuff can't break me," you've done the job.

7.2 Protecting Income and Assets

If your finances were a business, your paycheck would be the only machine in the factory that can't be replaced quickly. Your laptop can be replaced. Your car can be replaced. Even your savings can be rebuilt over time. But your ability to earn over the next twenty years—your *human capital*—is the engine that funds everything else. Protecting that engine isn't dramatic. It's practical. And it's the kind of boring decision that keeps a bad year from becoming a permanent detour.

Most people instinctively protect what they can see: the car, the phone, the couch. The twist is that your most valuable asset is invisible. It's future paychecks. When that income stops (injury, illness, lawsuit, death), the dominoes don't fall one at a time. They fall all at once: missed bills, drained emergency savings, retirement contributions paused, credit damage, strained relationships, and a creeping sense that you're working harder but getting nowhere.

Insurance can't prevent bad luck. It can prevent bad luck from turning into financial collapse.

The three threats that actually matter

There are a thousand niche policies you can buy. Only a few threats regularly turn ordinary households into case studies.

Income interruption: You can be healthy, careful, and responsible, and still get sideswiped by an injury, chronic condition, pregnancy complications, or a diagnosis that doesn't care about your calendar. If you need your paycheck to pay your mortgage, you have a single point of failure.

Death: Not as a philosophical event—as an economic one. If someone depends on your income (a partner, kids, aging parents, even a co-signed loan), your death creates a financial hole that grief doesn't fill.

Liability: The part of adulthood nobody puts on a vision board. A car accident, an incident on your property, a mistake that hurts someone, a moment of bad timing—and suddenly you're not arguing about a repair bill. You're negotiating with attorneys.

You don't need to insure every inconvenience. You need to build a fence around these three cliffs.

Disability insurance: the risk most people guess wrong

Ask ten people what insurance they *need* and you'll hear "health, auto, maybe life." Disability usually shows up as an afterthought, if at all. That's strange, because disability is the scenario where you're alive (so you still have expenses), but you can't reliably earn (so you don't have income). It's the worst possible combination for cash flow.

Disability insurance is simply income insurance. It replaces a portion of your paycheck if a sickness or injury keeps you from working. That's it. Not glamorous. Not fun to shop for. Also one of the clearest examples of paying a small, predictable cost to avoid a massive, unpredictable one.

A subtle point: this is about *long-term* disability, not a few weeks of inconvenience. Many employers offer short-term disability (often weeks to a few months). The financial catastrophe is long-term: months or years where your earning power is impaired while the rest of life keeps charging rent.

If you're building wealth, disability coverage can protect the exact phase when you're most vulnerable: before you've built enough assets to self-insure your lifestyle.

A real-world scenario: the ``good job" that wasn't protected

Consider a very ordinary profile: late 30s, two kids, a mortgage, a stable career, and just enough saved to feel like things are finally under control. Let's call him Marcus. He works in a role that's equal parts knowledge work and meetings—not heavy labor, not obviously dangerous. He exercises, doesn't smoke, and figures disability insurance is for people with riskier jobs.

Then a health issue shows up that doesn't look dramatic on TV but is devastating in real life: a neurological condition that makes concentration unreliable and fatigue unpredictable. Marcus can't perform consistently enough to keep his job. His employer's short-term disability runs out. The household tries to bridge the gap with emergency savings, then credit cards, then a 401(k) loan. The mortgage doesn't pause. Childcare doesn't pause. The kids' needs don't pause. Within a year, the financial stress becomes a second illness in the house.

This is the part people don't like to hear: a responsible person can do everything "right" and still be one medical event away from a cash flow emergency that lasts longer than their emergency fund.

Proper long-term disability coverage doesn't make the situation pleasant. It makes it *manageable*. It turns a free-fall into a controlled descent.

The policy details that change everything

Insurance contracts are made of boring words that decide whether you get paid when it counts. With disability insurance, the details worth caring about are the ones that determine *how easily you qualify* and *how long you're protected*.

Definition of disability: The phrase "own occupation" versus "any occupation" can be life-changing. In plain English, an "own occupation" definition is closer to: *if you can't do your job, you're disabled.* "Any occupation" leans toward: *if you can do any job at all, you're not disabled.* You can see how that might go for a surgeon with an injured hand, a pilot with vision issues, or a software engineer with cognitive impairment.

Elimination period: This is the waiting period before benefits start—often 90 days or more. The longer the waiting period, the cheaper the policy tends to be. It also determines how much cash you need to bridge the gap. The cleanest way to think about it: your emergency fund and your elimination period should be friends. If you have a 90-day waiting period, you want a cash buffer that can realistically cover those months.

Benefit amount and cap: Disability policies typically replace a percentage of your income, up to a limit. Don't assume "60%" means your lifestyle is covered; it means your pre-tax income is partially replaced, and the tax treatment depends on how premiums were paid (for example, employer-paid premiums may lead to taxable benefits). The practical move is to map your fixed bills and make sure the likely benefit covers the essentials.

Benefit duration: Some policies pay for a couple of years; others to a certain age. Shorter durations are cheaper, but they're also the ones most likely to run out while you're still living the same life.

You don't need to memorize the industry's vocabulary. You do need to insist on clarity: *What triggers benefits? How long until they start? How much do they pay? How long do they last?*

Life insurance: love, math, and avoiding expensive confusion

Life insurance is emotionally loaded because it forces a question nobody wants to linger on: *If I'm gone, who pays for the life I helped build?* The best life insurance decisions are made with compassion *and* a spreadsheet.

For wealth builders, the conversation usually lands on one big fork: term life versus whole life.

Term life is simple: you rent coverage for a set period (say 20 or 30 years). If you die during the term, it pays out. If you outlive the term, it ends. Clean, focused, and typically cost-effective when your goal is protection.

Whole life is more complex: it's permanent insurance with a cash value component and higher premiums. It's often marketed as an investment, a savings vehicle, or a "two-for-one" solution.

Here's the reality that cuts through the marketing: if your main goal is to protect your family while you build assets, term life is usually the tool that matches the job. It aims directly at the risk you're trying to transfer: *income disappearing before the plan is finished.*

Whole life can be appropriate in specific situations, but those situations are usually not "I'm trying to build wealth from zero." They're more like: estate planning needs, specialized tax considerations, or a deliberate strategy chosen with a fiduciary advisor who can explain tradeoffs without a commission-shaped grin.

If you take only one idea from the term-vs-whole debate, take this: if the premium is so high it crowds out your ability to fund emergency savings, pay down high-interest debt, or invest consistently, it's not protecting your wealth. It's consuming it.

How much term life is ``enough" without guessing

People commonly pick life insurance amounts the way they pick a hotel room: by vibe. The better approach is to insure the *obligations and dependencies* you leave behind.

A practical way to estimate a ballpark need is to add:

- Income replacement: the number of years your household would need your income (often until kids are grown or until a partner can realistically adjust), multiplied by the annual amount needed.
- Debt that wouldn't disappear: mortgage balance, private student loans, any co-signed obligations.
- One-time costs: final expenses, a buffer for immediate disruption, maybe childcare transition costs.
- Future goals you truly want funded even if you're gone (for some families, that includes education; for others, it's simply keeping the lights on).

Then subtract resources already available (savings, existing employer coverage, investments). What's left is the protection gap.

Two refinements make this more realistic:

- *Insure the years of vulnerability.* If you're 28 with toddlers and minimal savings, the gap is huge. If you're 55 with a paid-off house and a solid portfolio, the gap may be smaller.

- *Match the term to the timeline.* If your biggest risk window is "until the mortgage is paid" or "until retirement savings are built," your term should cover that.

This isn't about finding a perfect number. It's about avoiding the two classic mistakes: buying too little because it's uncomfortable to price reality, or buying too much because a salesperson anchored you to a scary scenario.

Employer life insurance: useful, but not a full plan

Many employers offer group life insurance, often as a multiple of salary. That's a nice base layer, but it has a trap: it's connected to your job. If you change employers, get laid off, or step away from work, that coverage can shrink or disappear right when your life is already in flux.

If someone truly depends on your income, consider employer coverage as the *bonus layer*, not the foundation. The foundation is a policy you control and can keep.

Liability insurance: the quiet shield that protects your future

Liability is the part of insurance people avoid thinking about because it feels accusatory, as if buying it means you're planning to hurt someone. In reality, it's the admission that accidents happen in a crowded world, and financial responsibility doesn't scale politely.

Liability coverage pays when you're legally responsible for injuring someone or damaging their property. It also typically pays for legal defense, which matters because legal defense can be expensive even when you did nothing intentionally wrong.

You can think of liability as the protection of *future you*. The you who has savings. The you who has investments. The you who finally has a little breathing room. Liability is the line between "a terrible incident" and "a terrible incident plus losing a decade of financial progress."

Auto and homeowners (or renters) policies include liability coverage, but the default limits are not always chosen with modern lawsuits and medical costs in mind. A serious accident can involve multiple injured parties, long rehab timelines, lost wages, and claims that climb quickly.

Umbrella insurance: the best bargain you hope you never use

An umbrella policy is exactly what it sounds like: extra liability coverage that sits on top of your auto and homeowners/renters liability limits. If a claim exceeds the underlying policy limits, the umbrella can provide additional coverage, often starting at $1 million.

Why it's so powerful is simple: big lawsuits don't respect your limits. They pursue damages based on the harm alleged and your ability to pay. If you have assets or high earning potential, you can look like a deeper pocket than you feel.

Umbrella coverage is often surprisingly affordable for the amount of protection it buys, but it comes with rules. Insurers typically require you to carry certain minimum liability limits on your auto and homeowners/renters policies before they'll issue the umbrella. And umbrellas don't cover everything; exclusions matter. The umbrella is not a magic shield for every kind of risk, and it doesn't cover damage to your own home or car.

The point isn't to be paranoid. It's to be appropriately defended.

A liability snapshot that makes it real

Imagine a common scenario: you're driving home, tired, it's raining, and traffic stops abruptly. You brake hard but still rear-end the car in front of you. That car hits the next car. Three cars, multiple injuries. One person can't work for months. Another needs surgery and physical therapy. Suddenly you're not arguing about a bumper. You're in the realm of medical bills, lost wages, and attorneys.

If your auto liability limit is $100,000 and the claim comes in at $600,000, the policy doesn't stretch. The remaining amount doesn't evaporate. That gap can target your savings and potentially your wages. If you have an umbrella policy that adds $1 million on top of robust underlying limits, the same terrible day can be financially survivable.

That's the difference between insurance as paperwork and insurance as a moat.

Protecting assets also means protecting access to cash

There's a simple reason financial emergencies feel like they happen in fast-forward: bills are due on a schedule, but good financial decisions take time. A market downturn is the worst moment to sell investments. High-interest debt is the worst bridge to use if you can avoid it. A strong cash buffer is not just "savings." It's *decision-making power.*

Consumer protection guidance consistently makes the point that emergency savings can reduce reliance on high-cost credit when life punches you in the mouth. When you combine that cash buffer with the right insurance, you create layers: small shocks are absorbed by cash, big shocks are transferred to insurance, and your long-term assets get to keep compounding instead of being liquidated under stress.

This is also where insurance choices connect back to your daily system. Higher deductibles and better liability limits often work best when you have an emergency fund sturdy enough to handle the deductible without drama.

A short note on ``protected" retirement accounts (and why it's not a free pass)

People sometimes hear that certain employer retirement plans have strong creditor protections under federal rules and assume they're basically lawsuit-proof. Reality is more nuanced. Some retirement assets may be protected in many situations, but there are exceptions, state law differences for certain account types, and specific carve-outs (like qualified domestic relations orders in divorce). The wrong takeaway is "I don't need liability coverage because my 401(k) is protected."

The right takeaway is calmer: legal protection for some assets is helpful, but it's not a substitute for managing liability risk in the first place. Lawsuits are stressful even when you're ultimately fine, and many households have assets outside retirement accounts anyway: home equity, taxable brokerage accounts, savings, and future wages.

Actionable takeaway: the 60-minute protection tune-up

Block an hour, once a year, and do a focused review that protects both your paycheck and your net worth.

- *Life insurance:* If anyone depends on your income, confirm you have term coverage you can keep independent of your employer. Write down the term length and the death benefit, and make sure the term covers your main vulnerability window.
- *Disability insurance:* Check whether you have long-term disability coverage, what percentage it replaces, the elimination period, and how disability is defined. If you're relying solely on employer coverage, price an individual policy as a comparison point.
- *Liability:* Look at your auto and homeowners/renters liability limits. If you have meaningful assets or rising income, get a quote for an umbrella policy and note the required underlying limits.
- *Cash alignment:* Make sure your emergency fund can comfortably cover your insurance deductibles and the

waiting period on disability coverage. If it can't, that's not a moral failure; it's a to-do list.

When the protection is right, you feel it in a strange way: fewer money arguments, fewer late-night "what if" spirals, and more consistent investing because you're not one unlucky event away from draining everything you've built.

7.3 Planning for the Long Run

Retirement planning gets marketed like a finish line: hit a number, quit your job, and start living the "real" life. The truth is messier and more interesting. The hard part isn't saving a pile of money. The hard part is converting that pile into paychecks for a few decades—without letting one bad stretch of markets turn your freedom into a part-time job you didn't want.

Think of it this way: while you're accumulating, time is your best employee. A market drop is annoying, but it often helps you because your contributions buy more shares. When you're drawing down, time becomes a stricter boss. You're no longer just investing; you're also *selling* to fund your life. The same market drop that once felt like a sale can become a trap if it hits at the wrong time.

That "wrong time" has a name: *Sequence of Returns Risk*. It's the reason two retirees can earn the same average return and end up in two completely different realities.

The 4% rule: useful, misunderstood, and still not magic

The famous "4% rule" is often repeated like a spell: withdraw 4% of your starting portfolio each year (adjusting for inflation), and you'll be fine. The reason it became popular is simple: it's a clear starting point. It gives your brain something to hold onto when the future feels like fog.

But it's not a guarantee, and it isn't really a rule. It's closer to an *estimate* based on historical simulations of diversified portfolios, meant to answer a very specific question: *What withdrawal rate had a high probability of lasting around 30 years in the past?*

Here's what people miss:

- The 4% idea assumes you keep investing during retirement; it's not "cash out and relax." It's "stay in the market, but don't take too much too fast."
- It's most sensitive to what happens early. The first decade can matter more than the last decade.
- Real retirement spending isn't a perfect inflation-adjusted line. It's lumpy: roofs, cars, medical costs, helping family, and sometimes a year of travel that's worth every penny.

A good long-run plan doesn't worship a number. It builds *guardrails*: a starting withdrawal rate, a method for adjusting, and a structure that reduces the chance you'll be forced to sell stocks at the worst possible moment.

Sequence of returns risk: same average return, opposite outcomes

Imagine two retirees with the same starting portfolio and the same average annual return over 30 years. If "average return" were the whole story, they should land in the same place. They don't—because the order of returns matters when you're withdrawing.

A simple way to feel it in your bones: if your portfolio drops early and you keep withdrawing to pay the bills, you're selling shares while they're cheap. That means fewer shares remain to participate in the eventual recovery. It's like cutting down fruit trees to survive winter; you might make it through the cold, but you've reduced your ability to produce fruit later.

Now flip the sequence. If you get strong returns early, the portfolio may grow enough that later downturns are just bruises, not broken bones. Withdrawals become a smaller percentage of a larger base. The plan has room to breathe.

It shows the same retiree in two universes: identical average returns, identical withdrawals, reverse order of returns. Only one of them gets to stay retired.

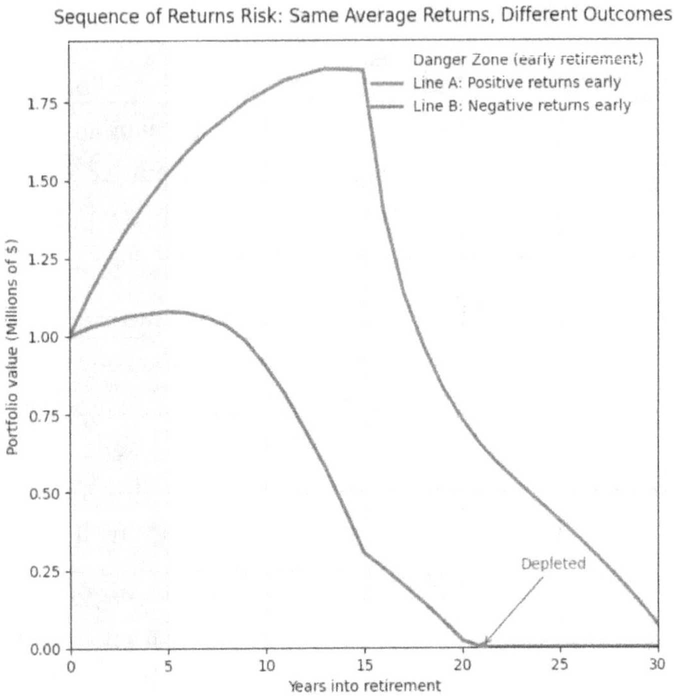

A concrete case: Elena retires into a bad first year

Elena is 62, newly retired, and proud of what she built: roughly $900,000 invested across retirement accounts and a taxable brokerage account. Her mortgage is paid off. Her spending plan is modest: $45,000 per year from investments until Social Security ramps up.

Her first year of retirement is also a year when the market drops hard. Not a mild dip—the kind that fills headlines and makes friends ask if you're "okay." Elena keeps withdrawing because the electric company doesn't accept "I'm waiting for a rebound" as payment.

Here's what happens psychologically, which is where sequence risk becomes real:

- In the accumulation phase, Elena used downturns to buy more. She felt brave.
- In the withdrawal phase, the downturn forces her to sell. She feels punished.
- She starts avoiding her account logins like it's a medical test result.

Elena's outcome depends less on whether markets eventually recover (they often do over long stretches) and more on whether she can avoid selling too much stock while prices are down. If she has a plan that creates a runway—cash, short-term bonds, or some other near-term funding source—she can keep paying herself without shredding the long-term engine.

The lesson is not "retire only in good years." The lesson is "structure your retirement so a bad first year doesn't own you."

The cash wedge (bucket strategy): buying time when markets are rude

A practical defense against sequence risk is a *cash wedge*, sometimes called a *bucket strategy*. The idea is straightforward: keep a designated pool of safer, more stable assets to fund near-term spending, so you're less likely to sell volatile investments after a drop.

One common layout looks like this:

- *Spending bucket:* cash and cash-like accounts for the next year (or two) of spending needs.
- *Stability bucket:* high-quality bonds or bond funds intended to cover the next several years.
- *Growth bucket:* diversified stock funds meant to power the decades.

This isn't about predicting markets. It's about controlling *behavior*. When markets fall, a bucket strategy gives you a script: "Spend from the spending bucket. Refill it later, when markets are calmer." It turns a terrifying decision (sell stocks now?) into a mechanical one (use the buffer you already set aside).

The bucket strategy has tradeoffs. Cash can lag inflation. Bonds can have their own bad years. But the goal isn't perfection. The goal is to reduce the probability that a downturn forces you to liquidate your long-term assets at a discount.

Withdrawal flexibility: the retirement superpower nobody brags about

If there's one lever that makes retirement plans more durable, it's the willingness to be a little flexible.

A rigid plan says: "I withdraw $X, inflation-adjusted, no matter what." A resilient plan says: "I start at $X, and I adjust within guardrails depending on markets, health, and real life."

Flexibility can be gentle:

- In down years, delay the big discretionary items (the kitchen remodel, the new car, the expensive trip).

- In up years, replenish cash reserves and maybe give yourself a controlled raise.
- Set a rule such as: if the portfolio is down more than a certain percentage, cap your "raise" or temporarily hold spending flat.

This isn't about deprivation; it's about staying in control. People imagine retirement risk as a single monster called "running out of money." In reality, it's a bunch of smaller monsters: a couple of bad market years, a roof replacement, a car problem, helping family, or unexpected healthcare costs. Small flexibility keeps those from stacking up into panic.

Health costs: the long-run variable that rarely stays polite

Even strong portfolios can get bullied by healthcare costs, especially later in life. Two facts matter for planning:

First, medical spending is uneven. Many retirees spend relatively little for years and then face a sharp increase due to chronic conditions, surgeries, or the need for assistance.

Second, *long-term care* is not the same thing as medical care. Long-term care is help with daily living—bathing, dressing, eating, supervision, and similar support. People often assume Medicare will cover it. Generally, Medicare does not cover *custodial* long-term care the way people imagine, which means families frequently pay out of pocket unless they qualify for Medicaid or have private coverage.

That gap is why "I'll just use my investments" can be both true and incomplete. You can use your investments, but your plan should acknowledge the possibility that an extended care need can act like a slow financial leak that becomes a flood.

Private long-term care insurance exists, but the market has changed over time, with fewer carriers and higher prices for many new policies. Some people choose coverage; others choose self-funding; many choose a hybrid approach (building extra assets, keeping home equity as a potential resource, and having candid family conversations early). The best plan is the one that doesn't pretend the risk is imaginary.

HSAs: a stealth retirement tool (with sharp edges)

If you're eligible for a Health Savings Account (HSA), it can be one of the most powerful long-run planning tools because it's designed around medical spending, which is a predictable category in retirement even if the timing is unpredictable.

An HSA is available only if you're covered by a qualifying high-deductible health plan (HDHP) and meet other eligibility rules. The attraction is the tax treatment: contributions can be tax-advantaged, growth can be tax-advantaged, and qualified medical withdrawals can be tax-free. Many people treat the HSA like a checking account for copays. A long-run planner may treat it more like a medical endowment: invest it, pay current costs out of pocket when feasible, and let the HSA compound for future healthcare needs.

But HSAs have rules that can surprise you. Eligibility can change mid-year. Contribution limits are annual and inflation-adjusted. There are timing rules (including a "last-month rule" with a testing period) that can create taxes and penalties if you contribute under an assumption that later becomes false. The point isn't to fear the HSA; it's to respect it. Used carefully, it's a rare financial tool that aligns perfectly with a major retirement expense category.

Turning the plan into something you can actually run

Long-run planning fails when it exists only as a number. "I need $1.2 million" is motivating, but it doesn't tell you what to *do* when markets wobble, when spending changes, or when life gets weird.

A runnable plan answers operational questions:

- Where does the next 12 months of spending come from?
- If markets drop 25%, which account gets tapped first?
- What spending gets paused first (without making life miserable)?
- When do you refill cash reserves, and from what assets?

You're essentially designing a household payroll system for the years when paychecks don't arrive from an employer. The more decisions you make in advance, the fewer you have to make during a downturn when your brain is not in the mood for spreadsheets.

Actionable takeaway: build a retirement ``shock absorber"

Write this on a page and keep it with your financial documents:

1. *Pick a starting withdrawal rate, not a lifetime promise.* Use a conservative starting point (the 4% idea is a common

baseline) and commit to annual review rather than blind faith.

2. *Create a cash wedge.* Hold 12–24 months of spending in cash/cash-like accounts, and a few additional years in a stability allocation you can draw from when stocks are down.
3. *Set two rules before markets test you:* (a) a simple trigger for tightening discretionary spending after a big drop, and (b) a trigger for refilling cash reserves after strong markets.
4. *Name your healthcare risk out loud.* Decide whether you're planning to self-fund long-term care risk with extra assets, exploring insurance, or combining strategies—but don't leave it as an unspoken assumption.

A retirement plan that can handle a bad early market isn't just mathematically stronger. It's emotionally calmer. And calm is a real asset when the headlines are loud and you still want your money to last.

7.4 Stress-Testing Your Plan

A plan that only works when life behaves is not a plan. It's a motivational poster with a spreadsheet attached.

The moment money gets real is the moment your assumptions get punched in the mouth: a layoff email on a Tuesday, a market drop that turns your "long-term investor" personality into a nervous insomniac, a transmission that dies the same week your kid needs braces. Most people don't fail financially because they didn't know they should save. They fail because they built a plan that required uninterrupted income, steady markets, and zero bad luck—and then felt ashamed when reality did what reality does.

Stress-testing is how you turn a fragile plan into a resilient one. It's not about predicting the next crisis. It's about rehearsing your response so you don't improvise with fear, pride, or a credit card.

The fire drill mindset: your future self will not be calm

Fire drills are useful precisely because nobody is calm during an actual fire. You don't run a drill because you expect the building to burn; you run it because you know that if it does, your brain will be loud and unhelpful. It will skip steps. It will bargain. It will panic-shop for solutions.

Money emergencies work the same way. In a crisis, people make the financial equivalent of running back into the building to grab a laptop: cashing out retirement accounts, taking predatory loans, selling investments in a downturn, or ignoring bills until the problem becomes a legal one. Stress-testing is how you pre-decide the boring, effective moves while your brain is still cooperative.

A good test asks two questions:

> What breaks first? What do I do first?

You're not looking for perfection. You're looking for a clean sequence of actions that keeps small damage from becoming permanent damage.

The three stress tests that reveal everything

Most financial "what-ifs" are just variations of a few core shocks:

Income shock: job loss, reduced hours, disability, commission drought, business slowdown.

Market shock: a 30–50% drop in stocks, or a multi-year period of disappointing returns right when you need liquidity.

Expense shock: medical bill, home repair, family emergency travel, legal fees, car replacement.

If you can survive those three, you can survive most of life. Not comfortably in every moment, but without burning down your future.

The weird part: people often try to stress-test by asking, "Do I have enough money?" The better question is operational: *Where does next month's money come from if X happens?* Stress-testing is logistics.

A concrete case: the six-month job loss that happens to ``responsible" people

Danielle and Priya are a dual-income household. They're not reckless. They have a budget, they invest in their 401(k)s, and they even have a small emergency fund. The plan assumes both incomes keep flowing. It also assumes that if something goes wrong, it will go wrong "a little."

Then Danielle's company restructures. She gets laid off with a small severance. It's February, which is a cruel time to job hunt in some industries because everyone is "waiting for budgets." Danielle expects to be back to work in two months. It takes six.

Here's what makes this story common: they didn't have a single catastrophic expense. They had a thousand small ones that didn't stop just because income did. Their emergency fund covered a chunk, but not the whole runway. They started floating groceries and utilities on a credit card. They told themselves it was temporary, which is exactly what everyone tells themselves at the start of a debt spiral.

They also made a classic mistake: they kept their lifestyle intact for too long because cutting expenses felt like admitting failure. By the time they got serious, the credit card balance had momentum.

A stress-tested plan would have changed the ending. Not by preventing the layoff, but by triggering an immediate protocol: cut discretionary spending in week one, preserve cash, negotiate bills early, and choose which assets to tap in a deliberate order rather than by desperation.

Build your own ``contingency triggers"

A contingency trigger is a pre-decided *if-then* rule that turns panic into procedure.

A few examples that actually work:

- *If income drops by 20% or more, then* pause all nonessential subscriptions and stop discretionary shopping within 48 hours.

- *If I'm unemployed for 30 days, then* I call lenders and request hardship options before I miss a payment.
- *If the market drops 30% and I'm still working, then* I do *not* reduce retirement contributions (unless cash flow truly requires it), because the contributions are my best "buy low" mechanism.
- *If my emergency fund falls below one month of expenses, then* no major purchases until it's rebuilt.

These are not moral statements. They're circuit breakers. They prevent you from running your financial system at unsafe temperatures.

The job-loss protocol: order matters

In a cash crunch, the order of operations can save you thousands of dollars and years of progress. People often do it backwards: they raid retirement accounts early and leave the easy wins untouched.

A clean job-loss protocol tends to follow a logic:

1. Reduce outflow immediately (cut the leak).
2. Use "free" or low-cost buffers first (severance, unemployment, cutting discretionary spend).
3. Use cash reserves next (that's what they're for).
4. Tap taxable assets before retirement accounts when possible (to avoid penalties and long-term damage), while being mindful of taxes.

5. Treat retirement accounts as last-resort resources because they have protections, tax advantages, and consequences when broken.

You don't need to follow that sequence blindly, and your situation may differ. But you do need *a* sequence. Without one, the decision gets made by stress—and stress prefers the fastest cash, not the smartest cash.

This is also where emergency savings earns its reputation. Consumer protection guidance consistently emphasizes that having cash on hand can prevent reliance on high-cost credit when shocks hit. That's not just good advice; it's a mechanical advantage. Cash buys you time. Time improves your decisions. Better decisions are cheaper.

Liquidity stress test: can you avoid selling long-term assets at the wrong time?

Here's a stress test that feels simple and turns out to be revealing:

> If I lost my income tomorrow, how many months could I cover *without* selling stocks?

This question forces you to separate two things that get lumped together: *wealth* and *liquidity*. You can be wealthy on paper and still be liquidity-poor. Home equity is real, but it's not a same-day solution. Retirement accounts are real, but cashing them out can be expensive and destructive. A portfolio is real, but selling into a downturn to pay rent is the kind of financial injury that leaves a scar.

A basic liquidity stress test includes:

- Cash and cash-like accounts (high-yield savings, money market funds).
- Expected severance and the realistic timing of it.
- Unemployment benefits (and the lag before they arrive).
- Bills that can be reduced quickly versus bills that are fixed.

If your answer is "less than one month," that's not a character flaw. It's information. It means your plan is depending on uninterrupted income, and your first upgrade is a larger emergency fund or a leaner fixed-expense structure—ideally both.

Market stress test: the year you retire is the year the market drops

Accumulation plans are often tested by boredom. Retirement plans get tested by timing.

A common nightmare scenario is simple: you retire, and stocks fall 40% in year one. Even if markets recover later, you're forced to withdraw from a smaller base, which magnifies the damage. This is the cousin of sequence of returns risk, but it can also hit people who aren't fully retired yet if they're planning a home purchase or a big life expense that depends on selling investments at a specific time.

The stress test looks like this:

> If my portfolio drops 40% next year, what changes in my life *immediately*?

A strong answer includes specific actions:

- Delay retirement by 6–24 months, if possible.
- Reduce withdrawals temporarily, especially discretionary spending.
- Use a cash wedge or short-term bond allocation to fund spending without selling stocks.
- Rebalance deliberately if your asset allocation drifted (which it will).

Notice the theme: the fix isn't "guess the market." The fix is having options.

Rebalancing: the discipline that feels wrong at the right time

Portfolio drift is normal. If stocks have a great run, they become a larger slice of your portfolio. If stocks crash, they become a smaller slice. Left alone, drift changes your risk level without asking permission.

Rebalancing is the act of restoring your target mix—selling a bit of what grew and buying what shrank, or redirecting new contributions to the underweight assets. Done well, it's a risk-control move, not a performance trick. It keeps your portfolio aligned with your actual tolerance for volatility.

The emotional challenge is that rebalancing often demands you do the opposite of your instincts:

- After a big stock run, rebalancing can mean trimming what feels like a winning streak.

- After a crash, rebalancing can mean buying what feels scary.

A stress-tested plan doesn't leave rebalancing to mood. It uses a rule: rebalance on a schedule (once or twice a year) and/or when allocations drift beyond a set band. Taxes and transaction costs matter, especially in taxable accounts, so the cleanest approach is often to rebalance with new contributions first, then make sales only when needed.

Target-date funds: same label, different engines

Target-date funds are popular because they promise simplicity: pick the year you plan to retire, and the fund gradually becomes more conservative over time. That simplicity is real. But so is the variation.

Funds with the same target year can have different "glide paths" (how quickly they reduce stock exposure), different underlying holdings, and different fees. Some are designed to reach their most conservative allocation *at* the target date (often called "to" retirement). Others keep shifting after the target date (often "through" retirement). Two funds can share a name like "2045" and behave like different animals.

The stress test isn't "Is a target-date fund good or bad?" It's: *Do I know what I own, and does it match my plan?* If your target-date fund still holds a lot of stocks when you expect to begin withdrawals, that can be perfectly reasonable—or it can be a surprise if you assumed it would be mostly bonds. Either way, surprises are expensive.

A simple practice: look up your target-date fund's current stock/bond mix, its glide path description, and its expense ratio. If you can't explain those three things in plain English, you're outsourcing more decision-making than you realize.

Expense shock test: your plan versus a $10,000 surprise

Not every crisis is a six-month saga. Some are quick and brutal. A sewer line collapses. A tooth breaks and becomes a root canal and crown. A family member needs a last-minute flight and support. The numbers vary; the pattern is the same.

Run this test:

> If I had to spend $10,000 in the next 30 days, where does it come from?

Answering "credit card" is not automatically wrong, but it is a sign. It means you're paying interest because your plan doesn't have shock absorbers. Sometimes that's temporary during the early wealth-building phase. Long-term, the goal is to graduate to a system where your first response is cash, not debt.

This is also where a high-deductible insurance strategy intersects with stress-testing. If you've chosen higher deductibles to keep premiums sane (a smart move for many), your emergency fund needs to be able to handle that deductible without a crisis. Otherwise you've built a plan that's theoretically efficient and practically fragile.

Make it real: write your one-page ``bad day" plan

The value of stress-testing isn't the thought exercise. It's the artifact: a short plan you can follow when you're tired, scared, or embarrassed.

A one-page "bad day" plan can include:

- The exact list of bills to protect first (housing, utilities, insurance, basic food, transportation).
- The first five expenses to cut (subscriptions, dining out, travel, shopping, upgrades).
- The order of cash sources to use (severance/unemployment, emergency fund, taxable assets, then last-resort options).
- The names and numbers you'll call (HR, lenders, insurance, a trusted family member).

It should feel almost too simple. That's the point. Complexity collapses under stress.

Actionable takeaway: schedule one ``fire drill" this month

Put a 45-minute appointment on your calendar and do a single stress test with numbers, not vibes.

1. Choose one scenario: six-month job loss, a 40% market drop, or a $10,000 surprise expense.
2. Write down your first five actions in order, including which accounts you will *not* touch unless absolutely necessary.

3. Define one trigger that forces action early (for example: "If I'm unemployed for 30 days, discretionary spending goes to zero").
4. Identify the weak link (too little cash, too much fixed spending, unclear investment mix, high-interest debt) and set one concrete fix with a date.

A strong plan doesn't eliminate emergencies. It makes them boring enough that you can handle them without breaking the life you're building.

7.5 Estate Planning Essentials

Estate planning sounds like something you do after you buy a boat you don't need. In reality, it's the instruction manual for your life in the moments when you can't speak for yourself—and the cleanup plan for the people you love when grief has already taken most of their bandwidth.

Without documents, the state provides defaults. Those defaults aren't evil; they're just generic. They don't know who you trust, who you're estranged from, who's good with paperwork, who's impulsive, who's quietly responsible, who will fight, or who will freeze. The law can distribute money. It can't distribute judgment.

The goal isn't to control the future from beyond the grave. The goal is to prevent avoidable chaos: frozen accounts, family conflict, expensive court processes, medical decisions made by the wrong person, and a situation where someone who's already hurting is forced to become a part-time paralegal.

The case study nobody plans for: ``We thought we had time"

Andre is 34. He and Maya have a toddler, a mortgage, and the kind of life that runs on shared calendars and coffee. Andre has a 401(k) through work and a small life insurance policy. They've talked about creating a will "once things slow down."

Then Andre has a medical emergency. He survives, but he's incapacitated for months. Suddenly, the biggest problems aren't emotional (though those are enormous). They're operational.

Maya can't easily access certain accounts because they're in Andre's name only. She can't sign certain documents on his behalf. Hospital staff ask who has legal authority to make decisions if a critical choice arises. Family members want to help, but they disagree on what Andre would want. Nothing is malicious; it's just uncertainty under pressure.

If Andre had died, the questions would multiply: who manages his estate, who receives what, how does the court process work, and what happens to the child if both parents were gone? Even with a loving family, the lack of paperwork is like removing street signs and expecting traffic to flow smoothly.

Estate planning doesn't remove grief. It removes bureaucracy from the path of grief.

The Big Three: Will, Power of Attorney, Medical Directive

If you do nothing else, do these three. They're not just for the wealthy. They're for adults who don't want their loved ones solving legal puzzles while sleep-deprived.

1) The Will: your distribution plan and your referee

A will is a legal document that states what happens to your property at death and who should administer your estate (often called an executor or personal representative). It can also nominate a guardian for minor children, which is one of the most important and emotionally loaded decisions you'll ever put on paper.

Two clarifying points keep people from making common mistakes:

A will doesn't control everything. Certain assets pass outside of a will by design: retirement accounts, many life insurance policies, and accounts with named beneficiaries or transfer-on-death registrations. This is where people get blindsided: they carefully write a will and later learn that a beneficiary form signed years ago overrides it for that asset.

A will isn't just "who gets what." It's also *who runs the process*. A good executor isn't necessarily the person you love most. It's the person who can handle paperwork, deadlines, conflict, and uncomfortable conversations with professionalism.

A practical way to choose an executor: pick the person who could coordinate a complicated move across state lines without losing receipts or patience. Then ask them if they're willing before you name them. Being named executor without warning can feel like being handed a second job during the worst week of your life.

2) Financial Power of Attorney: your legal stand-in while you're alive

A financial power of attorney (POA) allows someone (your agent) to act on your behalf for financial and legal matters if you can't. That can mean paying bills, dealing with insurers, managing bank accounts, filing taxes, handling a home sale, or interacting with employers and benefits.

The nightmare scenario isn't always death. It's incapacity. When someone is alive but unable to sign, speak, or reliably make decisions, the lack of a POA can force families into court proceedings to get authority. That's expensive, slow, and emotionally draining.

You're not just picking someone trustworthy. You're picking someone who can do the job under stress. Many people choose a spouse or partner; others choose a sibling, close friend, or professional fiduciary depending on family dynamics. The right answer is the person who won't misuse power and won't avoid using it when it's necessary.

3) Medical Directive: your healthcare voice when you can't talk

A medical directive (often called an advance directive) generally covers two things: your treatment preferences and who can make medical decisions for you (a healthcare proxy or agent). It exists for the exact moments when doctors need decisions and you can't provide them.

This isn't about predicting every medical scenario. It's about removing uncertainty. Many families don't fight because they disagree; they fight because they're terrified of making the wrong choice and living with it forever. A medical directive is a gift: it tells people what you want, and it names who should speak for you.

If you want your directive to work in real life, not just in theory, do two extra steps:

- Talk through your preferences with the person you appoint. A document without a conversation is a recipe for hesitation.
- Make sure it's accessible. If it's buried in a drawer, it won't help when you're in an ER.

The most common error: beneficiary forms quietly override your will

If estate planning were a video game, beneficiary designations would be the hidden trapdoor.

Retirement accounts (like 401(k)s and IRAs) and life insurance typically transfer to named beneficiaries. That transfer is often direct and doesn't require probate the way other assets might. It can also be fast, which is helpful for families who need liquidity.

But the same mechanism causes real-world disasters:

- An ex-spouse is still listed as beneficiary years after a divorce.

- A parent is listed, but the intention later became "my spouse and kids."
- "My estate" is listed, accidentally pulling the asset into probate and delaying access.
- Multiple accounts have inconsistent beneficiaries, creating fights that feel like betrayal but are really paperwork.

Some employer plans also have rules around spouses as beneficiaries unless a spouse waives rights. Whether you're married or single, the action item is the same: treat beneficiary forms as primary documents, not administrative trivia.

A simple habit: once a year, or after any major life change (marriage, divorce, birth, death, job change), log into every retirement account and insurance policy and confirm the beneficiaries. Screenshot or print the confirmation page and keep it with your records.

Transfer on Death (TOD): the underused tool that keeps things moving

Many non-retirement brokerage accounts can be registered with a transfer-on-death (TOD) designation. In plain English, TOD means: when you die, the assets transfer to the named person(s) without going through probate, while you keep full control while you're alive.

TOD can be a quiet hero for families. Probate isn't always awful, but it can be slow, public, and expensive. Anything that legally transfers outside of probate can reduce the administrative burden on survivors.

TOD isn't a substitute for thoughtful planning. It's a tool. The key is consistency: your TOD designations should match your overall intent, and they should be coordinated with your will and any trust planning if you have it.

Probate, taxes, and the executor's real job

People hear "executor" and imagine reading a will dramatically in a wood-paneled room. The real job looks more like managing a complex project with grief as a constant background noise.

Executors often need to:

- Collect documents, notify institutions, and manage account transfers.
- Pay valid debts and follow creditor priority rules.
- Keep records of transactions and distributions.
- File tax returns: the decedent's final income tax return (Form 1040), and potentially an estate income tax return (Form 1041) if the estate earns income during administration. Depending on circumstances, there may be additional filings.

This is where good planning acts like lubrication. Clean records, clear lists of accounts, up-to-date beneficiaries, and organized documents make the process dramatically easier. A messy financial life doesn't just inconvenience you; it offloads stress to the people you care about most.

One particularly practical move is to create an "In Case of Emergency" folder (digital or physical) containing:

- A list of accounts (banks, brokerages, retirement plans) and how to access them.
- Insurance policies and contact information.
- Mortgage, car loans, and other debts.
- Copies of your will, POA, and medical directive.
- Password manager access instructions (not the passwords themselves in a document that can be stolen).

It's not romantic. It's love in spreadsheet form.

Funeral planning: the consumer rights most people don't know they have

Funerals are expensive partly because they happen in a moment when families are vulnerable and time-compressed. The smartest financial move is not to be "cheap." It's to be informed.

In the U.S., consumers have specific rights when purchasing funeral goods and services. The Federal Trade Commission's Funeral Rule requires funeral providers to give itemized prices, disclose prices over the phone, and allows you to buy certain items (like caskets and urns) from third parties without being penalized. It also allows the use of an alternative container for cremation rather than requiring an expensive casket.

This matters because a funeral bill can easily become an unplanned financial shock on top of grief. A small amount of knowledge gives families the ability to slow down, compare options, and choose what matches their values instead of what matches a sales script.

A gentle but powerful planning step: talk with your closest person about your wishes. Not the full choreography, just the big decisions (burial versus cremation, religious components, preferred budget range). Clarity prevents overspending fueled by guilt.

Guardianship: the hardest paragraph you'll ever write

If you have minor children, you need to name a guardian in your will. Many parents avoid this because it feels like inviting the universe to consider a tragedy. But the choice doesn't create the risk. It creates the plan.

If you don't choose, a court may decide. The people you would have chosen might still end up as guardians, but the process can be contested, delayed, and emotionally brutal for family members who are already shattered.

A practical way to approach it:

- Choose someone whose values and stability you trust.
- Consider logistics: location, willingness, health, and their own family situation.
- Talk to them. A guardian nomination without consent is not a plan; it's a surprise.

Also consider separating roles: the person raising the kids doesn't have to be the same person managing the money. You can name different people for different responsibilities, depending on who is strongest in each role.

A simple table: what each document does

Document	*What it solves in plain English*
Will	Names who gets probate assets, who runs the process, and (if applicable) who should raise minor children.
Financial Power of Attorney	Lets someone handle your money/legal tasks if you're alive but unable to act.
Medical Directive / Healthcare Proxy	Names who speaks for you medically and clarifies treatment preferences when you can't communicate.
Beneficiary Designations	Directs where retirement accounts and life insurance proceeds go; often overrides your will for those assets.
Transfer on Death (TOD)	Moves eligible non-retirement brokerage assets to named beneficiaries without probate while you retain lifetime control.

Actionable takeaway: the weekend estate plan sprint

You don't need a perfect estate plan to get massive benefits. You need a *real* one.

Set aside a weekend block and do this:

1. *List your assets and accounts* (bank, brokerage, retirement, insurance, home) and note which have beneficiaries or TOD designations.
2. *Choose your three people:* an executor, a financial agent under POA, and a medical decision-maker. If you have minor children, choose a guardian and a backup.
3. *Create the Big Three documents* (will, financial POA, medical directive) using reputable legal help appropriate to your situation, and sign them correctly under your state's rules.
4. *Update beneficiaries* on every retirement account and life insurance policy so they match your intent.
5. *Write one page of practical instructions:* where documents are stored, who to call, and how to access key information.

When it's done, you'll feel a particular kind of relief: not the thrill of making money, but the calm of knowing that if life gets chaotic, the people you love won't have to guess what you wanted while also learning how to become you.

7.6 Building a Legacy of Generosity

Money is loud when it's scarce. Every decision feels like triage. But as your financial footing improves, a different question starts whispering in the background: *What is this for?* Not in a philosophical way—in a Tuesday-afternoon way, when you're paying bills and noticing that your life has become stable enough to look outward.

Generosity is often treated like dessert: nice if there's room, optional if there isn't. In practice, it's closer to physical therapy. It keeps your financial life from becoming rigid and self-centered. It protects you from a subtle kind of poverty: having a rising net worth and a shrinking sense of purpose.

The goal isn't to give until it hurts. The goal is to give in a way that makes you proud, makes your life feel coherent, and doesn't sabotage the foundations you're still building. The trick is turning generosity from an emotional impulse into a system.

The two kinds of giving (and why both can be wise)

There are two broad styles of giving, and most people lean hard into one without noticing.

Reactive giving happens when someone asks, a crisis appears, or a heartbreaking story finds you at the exact wrong time. It's human. It's often sincere. It's also a bit like grocery shopping while hungry: you spend more, you buy randomly, and you sometimes walk away feeling oddly unsatisfied.

Strategic giving is planned. It's a line item. It's a habit. It's giving with a map.

Strategic giving doesn't mean cold giving. It means you decide *in advance* what you care about, how much you can afford, and what a "yes" looks like. When the requests arrive (and they will), you're not making the decision with guilt in your throat. You're following a policy you wrote when you were calm.

There's a place for both. Reactive giving keeps you responsive to real people in your life. Strategic giving keeps you from being steered by whoever has the best marketing.

The real constraint: not money, but attention

People assume the limiting factor in generosity is dollars. Often, it's attention. The world has an endless supply of need, and your brain has a limited supply of emotional fuel. If you try to respond to everything, you'll either burn out or shut down.

A healthy giving plan does something counterintuitive: it makes peace with the fact that you can't help everyone. Then it focuses your help so it actually lands.

One simple way to decide what you're building is to answer two questions:

> Who do I feel responsible for? What problem do I want my money to push on?

Responsibility might mean your extended family and community. It might mean a cause that shaped your life. It might mean a problem you can't unsee: hunger, homelessness, education, domestic violence, animal welfare, medical research, climate resilience.

You're not choosing the *best* cause in some universal ranking. You're choosing the cause you're willing to support consistently, year after year, even when the news moves on.

A case study: the giving plan that survived a rough year

Tasha is a nurse. Her partner, Eli, runs a small business. They want to be generous, but their giving has always been chaotic: a few impulsive donations in December, a couple of sponsored walks, and occasional help for friends. Every year they end up with the same feeling: "We gave, but did it matter?"

They decide to make it systematic. They choose a modest baseline they can sustain: 2% of take-home pay, split into two buckets.

- *Local bucket (50%):* a food pantry and a domestic violence shelter in their county.
- *Impact bucket (50%):* a global health charity focused on malaria prevention.

They automate monthly donations. They also create a "spontaneous yes" fund of $500 per year for school fundraisers, friends' events, and the random asks that show up without warning. The fund has a rule: when it's gone, it's gone.

Then a tough year hits. Eli's business slows. Their income drops. Here's the surprising part: they don't stop giving entirely. They reduce giving temporarily to 1% and pause the spontaneous bucket. Because the plan is explicit, the change feels like adjusting a thermostat, not breaking a promise.

Tasha tells a friend later, "We didn't feel guilty. We felt consistent." That's the win. The system protected their finances *and* their values.

Effective altruism vs. values-aligned giving: a useful tension

There's a conversation in philanthropy that can make people defensive: effective altruism. In plain language, it means trying to do the most good per dollar, often using evidence and measurable outcomes. It asks questions like:

- How many lives are improved per $1,000?
- What interventions are proven to work?
- Are we funding the bottleneck or the brochure?

Then there's values-aligned giving, which prioritizes personal meaning, community ties, and identity:

- This shelter helped my cousin.
- This scholarship fund changed my life.
- This local program makes my neighborhood safer.

You don't have to choose one. The most satisfying giving plans often hold both:

- A portion for *meaning*: local, personal, relationship-based.
- A portion for *leverage*: high-impact interventions where the dollars stretch far.

This blend solves a real human problem. If you chase only measurable impact, giving can feel sterile. If you chase only emotional connection, giving can become scattered and less effective than it could be. A blended plan lets you be both compassionate and thoughtful.

The tax reality: giving is not a hack, but it can be optimized

It's tempting to treat tax benefits as a reason to give. That's backwards. The reason to give is values. Tax planning is the polishing step.

In the U.S., charitable contributions can be deductible if you donate to qualified organizations and you itemize deductions. Many households don't itemize every year. That doesn't mean giving is "wasted." It just means you shouldn't assume a tax break will show up.

Two practical implications:

- Give because you want to, not because you expect a refund to fund it.
- If you're near the line between itemizing and taking the standard deduction, you may be able to plan giving in a way that makes itemizing more likely in some years.

A classic tactic (when it fits your situation) is *bunching*: concentrating multiple years of giving into one year so the total deductions exceed the standard deduction, then taking the standard deduction in the off years. This is not about gaming the system; it's about aligning the timing of giving with how the tax rules work.

DAFs: a powerful tool with a public debate attached

A donor-advised fund (DAF) is a charitable account. You contribute money (or appreciated investments) to the DAF, potentially get a tax deduction in that year if you itemize, and then recommend grants to operating charities over time.

DAFs can be genuinely useful:

- They make bunching easier: you can fund the DAF in a high-income year, then give out grants gradually.
- They simplify recordkeeping: one receipt for the contribution, then a clean grant history.
- They can accept appreciated securities, which can make giving more tax-efficient than donating cash after selling.

They also sit inside a real public debate. Critics argue that because DAFs generally don't have the same required payout rules as private foundations, dollars can sit for a long time before reaching operating charities. Defenders argue that flexibility is the point: it allows thoughtful allocation, multi-year commitments, and stability during market cycles when nonprofits might otherwise get whiplash.

You don't need to pick a side to make a smart personal decision. You just need to be clear about your intent. If you open a DAF, write a personal rule like: *"I will grant at least X% of the balance each year,"* or *"I will grant all contributions within Y years."* The tool becomes aligned with generosity instead of becoming a holding pen for good intentions.

QCDs: giving directly from an IRA (for those who qualify)

For people who are older and have traditional IRAs, there's a giving method that can be especially clean: a qualified charitable distribution (QCD). In plain language, it allows eligible individuals to direct money from an IRA to a qualified charity in a way that can keep the distribution from being taxable income (within the rules). Because it reduces taxable income rather than creating an itemized deduction, it can be helpful even for people who don't itemize.

QCDs are rule-bound. They need to be executed correctly, and they have coordination issues (you generally can't treat the same dollars as both a non-taxable QCD and also take a charitable deduction for them). But conceptually, the appeal is elegant: instead of taking IRA income and then donating cash, you donate directly from the IRA.

This is one of those areas where a little procedural care matters more than big financial sophistication. If it's relevant to you, it's worth confirming the exact requirements and making sure the charity receives the distribution in the correct form.

Giving appreciated assets: the underrated move

Even without a DAF, one of the most tax-smart ways to give (when you have a taxable brokerage account) can be donating appreciated investments that you've held long enough to qualify for long-term capital gains treatment. The logic is simple: you avoid realizing capital gains on the appreciation, and the charity receives the value.

This isn't a loophole. It's an incentive embedded in the system to encourage giving. It also nudges you toward a smart portfolio habit: periodically reviewing concentrated winners and deciding whether some of that growth should fund your values.

If you're still early in your wealth-building and don't have taxable investments, ignore this without guilt. Your generosity can grow with your balance sheet.

The ``people in your life" problem: boundaries are part of generosity

The hardest giving decisions are rarely about charities. They're about people you know.

A friend needs help with rent. A sibling wants a loan. A cousin needs a car repair. You want to be generous, and you also don't want to become the family bank.

A giving system helps here too. It gives you language:

- "We have a set amount each year for personal help, and we've already allocated it."
- "We can help with groceries this month, but we can't take on ongoing bills."
- "We don't do loans; if we give, it's a gift. Here's what we can do."

Boundaries aren't selfish. They prevent resentment, which is the fastest way for generosity to die. If you give and then quietly stew, you didn't just lose money—you lost warmth.

Make it measurable without making it miserable

Some people resist measuring giving because it feels transactional. The point isn't to reduce compassion to a spreadsheet. The point is to ensure the giving you do is the giving you *meant* to do.

Two measurements are surprisingly effective:

- *Giving rate:* a percentage of income (even a small one). This scales with your life and avoids the awkwardness of choosing an arbitrary dollar amount every year.
- *Focus list:* your top 3–5 recipients or causes. Not as a prison—as a default.

If you want extra rigor, you can add a "review day" once per year: read short annual reports, check basic transparency indicators, and confirm you still believe in the mission. Then you're done. You don't need to turn giving into homework.

Then automate the baseline giving monthly. The automation is not laziness; it's loyalty to your own values. It turns generosity into something you do on purpose, not something you do only when you feel guilty enough to click "donate."

www.ingramcontent.com/pod-product-compliance
Lightning Source LLC
LaVergne TN
LVHW052024200926
848481LV00005B/1575